PHOTOBIOLOGY OF THE SKIN AND EYE

DRUG AND CHEMICAL TOXICOLOGY

Series Editors

Frederick J. DiCarlo
Science Director
Xenobiotics Inc.
Denville, New Jersey

Frederick W. Oehme
Professor of Toxicology, Medicine and Physiology
Director, Comparative Toxicology Laboratories
Kansas State University
Manhattan, Kansas

Volume 1 Toxicity Screening Procedures Using Bacterial Systems, *edited by Dickson Liu and Bernard J. Dutka*

Volume 2 Chemically Induced Birth Defects, *James L. Schardein*

Volume 3 Neurotoxicology, *edited by Kenneth Blum and Luigi Manzo*

Volume 4 Statistics for Toxicologists, *David S. Salsburg*

Volume 5 Photobiology of the Skin and Eye, *edited by Edward M. Jackson*

Other Volumes in Preparation

PHOTOBIOLOGY OF THE SKIN AND EYE

edited by

Edward M. Jackson, Ph.D.
Noxell Corporation
Hunt Valley, Maryland

Marcel Dekker, Inc. **New York • Basel**

Library of Congress Cataloging in Publication Data

RL803.P49 1986 616.5 86-8835
ISBN 0-8247-7624-0

Photobiology of the skin and eye.

(Drug and chemical toxicology ; v. 5)
Includes bibliographies and index.
1. Skin—Effect of radiation on. 2. Eye—Effect of radiation on. 3. Light—Physiological effect. 4. Photosensitization, Biological. 5. Photobiology.
I. Jackson, Edward M., [date] II. Series: Drug and chemical toxicology (New York, N.Y.: 1984); v. 5.
[DNLM: 1. Eye—radiation effects. 2. Light—adverse effects. 3. Photochemotherapy. 4. Photosensitivity Disorders—etiology. 5. Skin—radiation effects. 6. Sunlight—adverse effects. WI DR513F v. 5 / WR 160 P574]

MARCEL DEKKER, INC.
270 Madison Avenue, New York, New York 10016

Current printing (last digit):
10 9 8 7 6 5 4 3 2 1

PRINTED IN THE UNITED STATES OF AMERICA

ABOUT THE SERIES

Toxicology has come a long way since the ancient use of botanical fluids to eliminate personal and political enemies. While such means are still employed (often with more potent and subtle materials), toxicology has left the boiling-pots-and-vapors atmosphere of the "old days" and evolved into a discipline that is at the forefront of science. In this process, present-day toxicologists adopted a variety of techniques from other scientific areas and developed new skills unique to the questions asked and the studies being pursued. More often than not, the questions asked have never been thought about before, and only through the advances made in other disciplines (for example, in analytical chemistry) were the needs for answers raised. The compounding concerns of society for public safety, the maintenance of environmental health, and the improvement of the welfare of research animals have expanded the boundaries in which toxicologists work. At the same time, society has spotlighted toxicology as the science that will offer the hope of safety guarantees, or at least minimal and acceptable risks, in our everyday chemical encounters.

This *Drug and Chemical Toxicology* series was established to provide a means by which leading scientists may document and communicate important information in the rapidly growing arena of toxicology. Providing relevant and forward-looking subjects with diverse and flexible themes in an expedited and prompt publication format will be our goal. We will strive in this vehicle to provide fellow toxicologists and other knowledgeable and interested parties with appropriate new information that can be promptly applied to help answer current questions.

The study of the effect of light upon biological systems (photobiology) is a unique aspect of toxicology. It is also a developing field that is attracting more attention and concern—from the increasing social attraction of ultraviolet light exposure and skin tanning with the associated documented skin cancer risks, to the correlation of exposure to sunlight with heightened risk of cataracts in the increasing aged population. Dr. Jackson's enlightening compilation of chapters on this speciality introduces photobiology to us in its full range of ultraviolet impacts. Reviews of the scope, principles, test methodologies, and dermal immune responses are followed by discussions of photocarcinogenesis, regulatory issues in marketing skin protection, and the therapeutic use of ultraviolet radiation in medicine. The final chapter shifts to the eye and emphasizes the harmful effects of light exposure to that vital organ.

The inference is clear: The gift of sunlight is a mixed blessing; beware to those who abuse it! This fifth volume in the *Drug and Chemical Toxicology* series offers a spotlight in which the exciting and challenging discipline of photobiology is introduced to our readers. Get acquainted!

Frederick W. Oehme
Frederick J. DiCarlo

PREFACE

Photobiology of the Skin and Eye is the result of two years of planning and one year of research and writing by the distinguished group of authors, each a recognized expert in his field. The purpose of this book is to provide the reader with a single authoritative source of up-to-date photobiological information and references.

Photobiology as a scientific discipline currently encompasses over a dozen specialties and has its own international society, The Society of Photobiology (1340 Old Chain Bridge Road, Suite 300, McLean, Virginia 22101), founded in 1972. Photobiology is already one of the fastest growing fields in science and medicine.

From the various subspecialties in photobiology, we have chosen seven as topics for chapters in this book: an overview of the field of photobiology (Megaw and Drake), phototoxicity (Gabriel, Forbes, and Davies), photoimmunology and photoallergy (Stephens and Bergstresser), photocarcinogenesis (Forbes, Davies, Urbach, and Cole), sun protection factor testing (Morton and Murphy), photochemotherapy (Berger and Reilly), and photoeffects on the eye (Dayhaw-Barker and Barker).

At first glance, it might seem odd to limit the scope of a book on photobiology to the skin and eye. However, no topic could be more appropriate for a book on photobiology. One reason is that the skin and the eye are the only two human organs directly exposed to light. In fact, it is surprising that the application of light and the study of the effects of light on the skin—our interface with the environment—and the eye—the central nervous system's window on the outside world—are so late in developing into a field of specialization. Further, more similarities than dissimilarities exist between the skin and the eye in reacting to light than we immediately recognize. For example, the pigment melanin is produced by the epidermal melanocytes packaged into melanosomes and injected into the surrounding keratinocytes in response to exposure to light. In the skin, melanin serves a protective purpose. However, in the retina of the eye, melanin is resident and reacts to light. Here melanin serves a functional purpose.

The first chapter is an overview of the physical and chemical principles of photobiology. This chapter quite properly begins with a fundamental fact: light effects can be beneficial or detrimental, depending on the circumstances. The second chapter reviews the three types of phototoxic (photoirritant) cutaneous reactions in photosensitization. Human, animal, and *in vitro* phototoxicity test methodologies are described. The third chapter is a discussion of photoimmunology and photoallergic skin responses. A description of preclinical and clinical test methods and models for determining photoallergic contact dermatitis concludes this chapter. In the chapter on photocarcinogenesis, the theory of photocarcinogenesis is discussed from data generated with the current animal models used to understand this deleterious effect of light exposure. We currently have available consumer products containing chemical sunscreen agents, so the next chapter considers the regulatory status and methods for determining sun protection factors. Then a chapter on the use of light in medicine, the field of chemotherapy, describes photochemotherapies for psoriasis and hyperbilirubinemia and the diagnostic tool of hematoporphyrin derivative. Finally, the concluding chapter on photoeffects on the eye describes the deleterious ocular effects from light exposure and the clinical diagnostic tests and tools available to the scientist and physician.

We are now at a developmental apex in the field of photobiology. Differing scientific and medical terminology in the fields of toxicology, dermatology, and ophthalmology is becoming more uniform, and certain principles may be evolving. For instance, UVA appears to be primarily responsible for phototoxic and photoallergic reactions, while UVB appears as the principal band of light resulting in photocarcinogenic effects, at least in animal models. On the other hand, UVA is currently a primary diagnostic light band.

We are on the threshold of exciting developments in our continuing understanding of both the beneficial and deleterious effects of light. At the same time, we are developing exquisite diagnostic and prognostic tools to further our knowledge of photobiology.

Edward M. Jackson

CONTENTS

CONTRIBUTORS

Felix M. Barker II, O.D. The Pennsylvania College of Optometry, Philadelphia, Pennsylvania

Richard S. Berger, M.D. Division of Dermatology, University of Medicine & Dentistry of New Jersey-Rutgers Medical School, New Brunswick, New Jersey

Paul R. Bergstresser, M.D. Department of Dermatology, University of Texas Health Science Center, Dallas, Texas

Curtis Cole, M.S. Center for Photobiology, Skin and Cancer Hospital, Temple University Health Sciences Center, Philadelphia Pennsylvania

Ronald E. Davies, Ph.D. Center for Photobiology, Skin and Cancer Hospital, Temple University Health Sciences Center, Philadelphia, Pennsylvania

Pierrette Dayhaw-Barker, Ph.D. The Pennsylvania College of Optometry, Philadelphia, Pennsylvania

Lynn A. Drake, M.D. Department of Dermatology, Emory University, and Veterans Administration Medical Center, Atlanta, Georgia

P. Donald Forbes, Ph.D. Center for Photobiology, Skin and Cancer Hospital, Temple University Health Sciences Center, Philadelphia, Pennsylvania

Karl L. Gabriel, V.M.D., Ph.D. Biosearch, Incorporated, Philadelphia, Pennsylvania

Judith M. Megaw, Ph.D. Department of Ophthalmology, Emory University, Atlanta, Georgia

Joseph J. P. Morton, Ph.D. Gillette Medical Evaluation Laboratories, Rockville, Maryland

Emalee G. Murphy, Esq. Cosmetic, Toiletry and Fragrance Association, Inc., Washington, D.C.

George D. Reilly, M.D. Division of Dermatology, University of Medicine & Dentistry of New Jersey-Rutgers Medical School, New Brunswick, New Jersey

Thomas J. Stephens, Ph.D. Mary Kay Cosmetics, Inc., Dallas, Texas

Frederick Urbach, M.D. Center for Photobiology, Skin and Cancer Hospital, Temple University Health Sciences Center, Philadelphia, Pennsylvania

PHOTOBIOLOGY OF THE SKIN AND EYE

1

PHOTOBIOLOGY: AN OVERVIEW

*JUDITH M. MEGAW, Ph.D.**
Department of Ophthalmology
Emory University
Atlanta, Georgia

LYNN A. DRAKE, M.D.
Department of Dermatology
Emory University
and
Veterans Administration Medical Center
Atlanta, Georgia

Introduction

Without the sun, life as we know it could not exist on earth. Solar energy (electromagnetic radiation) provides both the warmth required to make the planet habitable and the primary energy source for living organisms. How solar energy might influence and interact with living things has been of interest since ancient times. Today, the influences and interactions of nonionizing electromagnetic radiation, both from the sun and man-made sources, on biological systems, and the use of such radiation to study biological processes constitute the purview of the science of photobiology. Areas of interest falling under this general heading include vision, photosynthesis, environmental photobiology, bioluminescence, photomedicine, and others.

In man, nonionizing electromagnetic radiation can have very beneficial effects, such as the enhancement by ultraviolet radiation of the conversion in the skin of 7-dehydrocholesterol to vitamin D_3. On the other hand, exposure to solar radiation and man-made sources of electromagnetic radiation can have a number of deleterious effects, causing or aggravating a variety of processes related to aging and pathologic conditions. The skin and eye are the only organs that receive direct exposure to such radiation. The present issue, therefore, reviews what is presently known about the etiologic or contributory role of nonionizing electromagnetic radiation (wavelengths 200–800 nm) in these tissues with respect to toxic, immunologic, allergic, and carcinogenic effects. In addition, applications of photochemotherapeutic modalities and measures to prevent deleterious effects of exposure to nonionizing electromagnetic radiation will be discussed. This paper is intended to provide an overview of photobiology in general. For detailed discussion of each of the above subjects, the reader is referred to the articles that follow in this volume.

Basic Principles

Electromagnetic Radiation

Certain principles must be grasped in order to appreciate fully the responses of biological systems to exposure to radiant energy. The most basic of these is an understanding of the characteristics and behavior of electromagnetic radiation. The electromagnetic spectrum consists of those forms of energy that can be transmitted through a vacuum. Depending upon the interactions between electromagnetic radiation and matter being examined, such radiation may be described either in terms of its wave properties (wave description) or in terms of quantum theory (particle description). In the former, electromagnetic energy can be viewed as continuous waves produced by rapid oscillations of electric and magnetic fields perpendicular both to each other and to the direction of the wave propagation. When considered as a wave, such energy can be specifically characterized in terms of its wavelength. The unit of measure for wavelength is the meter or fraction thereof (e.g., nanometer) and represents the physical distance between one point in an oscillation (or wave) and the next exactly equivalent point in that oscillation. The wavelength of electromagnetic radiation bears an inverse relationship to its frequency: the number of complete oscillations passing a point in space per unit time. Einstein postulated that the maximal speed achievable for radiant energy was a constant equal to the speed of light in a vacuum, and that for any form of electromagnetic radiation, the product of the wavelength (λ) and its frequency (ν) was equal to this constant: $\lambda \times \nu = C = 3 \times 10^8$ m/sec.

Scattering, reflection, and refraction of electromagnetic radiation are best understood by looking at the radiation as continuous waves. The wave description, however, is inadequate for explaining such phenomena as absorption and emission of light by molecules. Such phenomena require that electromagnetic radiation be described in other terms.

Quantum theory tells us that electromagnetic radiation is propagated in discrete packets called photons. The photon energy of a given electromagnetic radiation depends upon its frequency (and wavelength). The higher the frequency (or the shorter the wavelength) of the osçillations occurring in a given energy wave, the greater the energy possessed by that wave. Planck's equation defines this relationship thus: The energy of a given photon is equal to the product of the frequency of the energy wave and a constant, h. This constant, equal to 6.624×10^{-34} J-sec, is called Planck's constant. Put another way, the energy of a given photon is equal to the product of h and the speed of light, c, divided by the wavelengh of the radiation:

$$E = h\nu = \frac{hc}{\lambda}$$

Electromagnetic radiation is classified according to wavelength ranges. The classification is based in part on the effects of the various energy levels of radiation on molecules and in part on the means required for their production and detection. As noted earlier, the energy of electromagnetic radiation is a direct function of its frequency and inversely related to its wavelength. If the energy of a photon of electromagnetic radiation exceeds the bond energies of molecules exposed to it, ionization and bond breakage can occur. Thus the short-wavelength, high-energy forms of electromagnetic radiation (e.g., γ and x rays) are termed ionizing radiation. Longer-wavelength, lower-energy radiation ($>$ 800 nm) will cause an increase in molecular rotation and vibration, which is a thermal effect. On the other hand, wavelengths between 200 and 800 nm are capable of causing chemical changes if absorbed by molecules. These changes are a consequence of alterations in the energies of electrons within the molecule. In some cases, such excitation actually may be sufficient to break chemical bonds, while in other cases the energy that excites electrons to higher energy states is dissipated in other ways.

The prefix for photobiology, "photo," derives from the Greek "*phos*," meaning light and, in its strictest definition, implies only that radiation to which the retina (human) is sensitive in terms of vision, that is, visible light. In actual practice, the definition of "light" as applied to photobiology includes ultraviolet as well as visible and near-infrared electromagnetic radiation (wavelengths 200–800 nm). These limits are chosen principally because, as noted above, these wavelengths can produce chemical changes when they are absorbed by molecules.

The sun is the main ubiquitous environmental source of electromagnetic radiation to which man is exposed. Only a very small portion of the radiant energy from the sun reaches the earth. In addition, the spectral distribution and intensity of solar electromagnetic radiation actually reaching the earth's surface vary with time and location. Absorption by the earth's atmospheric ozone layer and molecular scattering of shorter wavelengths entering the atmosphere serve to attenuate visible light reaching the earth's surface, and effectively remove wavelengths of solar radiation shorter than 290 nm. Water vapor and CO_2 in the atmosphere further serve to modify the spectral distribution of solar energy reaching the earth by absorbing red and infrared radiation. The electromagnetic radiation that does reach the earth's surface from the sun contains wavelengths from about 290 to 4,000 nm. This radiation is subcategorized as UV, visible, and infrared depending on its wavelength.

The UV portion of the electromagnetic spectrum may be further subdivided into UVC, UVB, and UVA radiation. This classification derives in part from the reactions evidenced by skin upon exposure to the various wavelength ranges. UVC radiation coming from the sun, consisting of wavelengths 200–290 nm, is effectively screened out by the ozone layer of the atmosphere and does not reach the earth's surface; however, man-made sources for these wavelengths have been developed (e.g., the low-pressure mercury germicidal lamp). UVC can produce skin erythema in moderate doses, but such erythema is not usually related to subsequent skin pigmentation. UVB (290–320 nm) and UVA (320–400 nm) radiation from the sun does reach the earth's surface from the sun. UVB, also called sunburn radiation, produces erythema in skin that usually develops later than erythema produced by UVC radiation. In addition, UVB-induced erythema is usually followed by increased skin pigmentation in the exposed areas. UVA in moderate doses does not produce skin erythema. This wavelength range, also called blacklight or "near" UV, is important, however, in a number of photosensitized skin reactions.

Wavelengths between 400 and 700 nm constitute the visible portion of the electromagnetic spectrum, beginning with violet light (400–424 nm) and ending with red (647–700 nm). The remainder of solar electromagnetic radiation reaching the earth is primarily in the infrared region of the spectrum (> 700 nm).

When considering the potential biological effects of electromagnetic radiation, one must remember that we have developed our own means for producing most forms. While many have significant therapeutic and experimental value, potential hazards may arise from improper use or accidental exposure to these sources.

Energy Absorption and Excited States of Molecules

To understand mechanisms extant in responses of biological systems to light, one must remember that light must be absorbed in order for a photochemical event to take place. This is the First Law of Photochemistry and was propounded by Grotthus and Draper in 1818. For absorption to occur, some interaction must take place between the incident energy and the absorbing molecule.

In molecules, the distribution of energy possessed by the electrons is not continuous. The laws of quantum mechanics tell us that for a given molecule only certain spatial distributions around the nuclei are "allowable" for the electrons. Each of the specific allowable distributions corresponds to a particular electronic state of the molecule, which in turn has a specific molecular energy. Again, from quantum mechanics, only certain energies are "allowed" for a given molecule. As a consequence, differences in energies of electronic states have fixed gaps. For a given molecule, the electronic state corresponding to the lowest energy is called the ground state. If the molecule is in an electronic state that has higher energy than the ground state, the molecule is said to be in an excited state. In addition to fixed electronic energy levels, for a given molecule, a set of molecular vibrational modes exists for each electronic state. Moreover, there is a set of fixed energy levels associated with each vibrational mode, the spacings of which depend upon the frequency of the vibration. Compared to spacings between electronic energy levels, the vibrational energy spacings are small. In order for a molecule to undergo a transition from one electronic state to a higher energy electronic state, energy must be absorbed. For absorption of electromagnetic energy to occur, the law of conservation of energy must be satisfied in the process: the energy absorbed must be equal to the energy gained by the molecule. Since absorption of photons of electromagnetic energy is an all-or-none process (i.e., only whole photons can be absorbed), in order for a molecule to absorb electromagnetic radiation, some combination of electronic state and vibrational level of the molecule must exist, higher in energy by an amount exactly equal to the photon energy of the radiation.

In the absorption process, the basic interaction between a photon of electromagnetic energy and a molecule is between the electric field of the light and the electrons of the molecule. As noted before, the electrons of a molecule occupy specific regions, called orbitals, in space around the nuclei. Each orbital can be occupied by, at most, two electrons and the two electrons within a given orbital, according to the Pauli exclusion principle, must have opposite spins. The spins of electrons in a molecule can be utilized to calculate the state multiplicity of the molecule. The state multiplicity $= 2S + 1$, where S is the absolute value of the total electron spin, which is determined by assigning values of $+\frac{1}{2}$ and $-\frac{1}{2}$ to the two opposite spins of electrons. Where there are equal numbers of electrons having + and − spins, $S = 0$ and the state multiplicity $= 1$. This is called a singlet state. In organic molecules, a state multiplicity of 3 can also be found. This is termed the triplet state. Most molecules in the ground state are also singlet-state molecules.

When a photon of electromagnetic energy is absorbed by an electron in a singlet-state molecule, the electron will be promoted to the next higher empty orbital. In the initial absorption process, the spin of the promoted electron remains unchanged, and the excited state of the molecule first generated is also a singlet. The Pauli principle, however, allows either spin for unpaired electrons in orbitals, thus in one-electron excited states, if the spin of the electron becomes reversed after promotion to a higher energy level, a total spin of 1 may be obtained, resulting in a triplet excited state. Because the electron–electron repulsion is less, the triplet state of a given orbital configuration always is lower in energy than the corresponding singlet state. (The significance of the excited singlet and triplet states of molecules in photochemistry will be addressed in more detail below.)

From the foregoing discussion, it is clear that molecules of a given type can only absorb electromagnetic energy of specific photon energies or wavelengths. The probability that a given wavelength will be absorbed depends on the energy separation between electronic states, nuclear positions, and other factors. This probability can be expressed by the molar extinction coefficient, ϵ, of the absorbing medium. Empirically, the absorbance, A, of a solution, can be measured by exposing the solution to a beam of light and comparing the intensity of the light before it enters the solution with the intensity of the light that passes through the solution. If the incident light is monochromatic (i.e., composed only of one wavelength), the absorbing medium is homogeneous, and the absorbing centers act independently of one another, the light absorbed will be related to the molar extinction coefficient according to the Beer-Lambert Law:

$$A = \log_{10}(Io/I) = \epsilon cl$$

where A = absorbance, Io = the intensity of the light before it enters the medium, I = the intensity after it has traversed the medium, c = the concentration of the solution, and l = the path length traversed by the light through the absorbing medium. If c and l are kept constant, a plot of the probability of a given medium for absorbing energy of different wavelengths can be achieved by varying the wavelength and measuring the absorbance at each wavelength. This yields the absorption spectrum for a given medium. For atoms, the absorption spectrum will consist of only a few lines, reflecting the small number of electronic states extant for the atomic species. When atoms are covalently bonded in molecules, the additional vibrational and rotational levels associated with electronic states increases, resulting in absorption occurring at many wavelengths. The lines produced in the absorption spectrum are therefore greater in number and so close together that they cannot be distinguished from one another. Thus, only a profile of the line heights is recorded. This profile consists of peaks and valleys. The peaks are termed absorption maxima.

Based on the analyses of large numbers of organic compounds, empirical rules have been derived which relate a compound's molecular structure to the nature of its absorption spectrum. Fully saturated groups absorb below 200 nm. On the other hand, compounds having unsaturated groups or conjugated combinations of such groups will absorb in the UV and sometimes the visible region of the electromagnetic spectrum. (The absorption by a compound of wavelengths in the visible region will confer a characteristic color upon it; e.g., rhodopsin, a visual pigment, absorbs in the blue–green region and transmits or reflects all red light, hence it appears red.) The absorption of longer wavelengths of light in molecules with unsaturated groups arises as a consequence of the orbital configurations related to the unsaturated bonds. The energy gaps between the highest energy orbitals containing electrons available for excitation and the lowest energy level orbitals to which they may be promoted decrease because the electrons are less tightly held by their nuclei. In general, increases in the number of conjugated double bonds within a compound results in shifting of the absorption peaks detected for the compound to higher wavelengths. For example, in ethylene, which has only one double bond, more energy is required to generate the first excited state than is required in benzene which has three double bonds, thus ethylene absorbs light at 200 nm, while benzene absorbs at 280 nm. While in smaller molecules the number of conjugated bonds present is directly proportional to the absorption maximum observed, in biological macromolecules such as nucleic acids and larger proteins, absorption maxima are observed which are lower or higher than would be predicted on the basis of molecular structure. This phenomenon is termed hypo- or hyperchromicitiy.

The unsaturated absorbing center of a molecule, comprised of all the conjugated multiply bonded atoms within the molecule, is termed a chromophore. In addition to the number of unsaturated bonds present, other factors also influence the absorption characteristics observed for a given compound. The addition of charged side groups can produce large changes in the absorption spectrum of a chromophore. In compounds containing ionizable size groups, shifts in observed absorption maxima can occur with changing pH. In addition, since the temperature affects the vibrational state of molecules as well as the degree of Brownian motion (and hence the frequency of molecular collisions and the duration of excited states), the absorption spectrum derived for a compound is also dependent upon the temperature at which absorption is measured. All these characteristics are important in assessing the potential phototoxic or photosensitizing properties of agents in biological systems, as we shall see later.

Relaxation of Excited-State Molecules

As noted earlier, the first law of photochemistry states that light must be absorbed for a photochemical event to occur. We have further established that only wavelengths of specific photon energies can be absorbed by a given molecule. When a photon is absorbed by a molecule in the ground state, the molecule undergoes transition to an excited singlet state. The excited singlet-state molecule can rid itself of the excess energy acquired in the absorption process by a number of pathways. The pathway followed depends upon the distribution of the electrons in the excited state molecule, the availability of other molecules with which to interact, and other factors. One pathway by which the excited singlet-state molecule can return to the ground state is by emission of light (fluorescence). If the excited singlet and the ground singlet state have almost the same arrangement of nuclei, the fluorescence emission spectrum and the absorption spectrum will overlap and will often be mirror images of each other. In the excited singlet-state molecule, the promoted electron can undergo a spin "flip," resulting in lowering of the energy of the excited molecule. This transition alters the spin multiplicity of the molecule, producing an excited triplet state. Excited singlet states of molecules have lifetimes of from 10^{-9} to 10^{-5} sec. The lifetimes of excited triplets are much longer, 10^{-3} to 10^{-2} sec, making them available longer for molecular interactions. Excited triplets can also undergo radiative decay to the ground state. In this case, the light emission is called phosphorescence.

Radiative decay processes are not the only means by which excited-state molecules can discharge excess energy. As already noted, conversion of a molecule from one excited state to a lower energy state of different multiplicity can also occur. Such a conversion is termed intersystem crossing. When intersystem crossing occurs, the energy lost by the excited electron appears as vibrational energy in the molecule. This vibrational energy is rapidly lost to the surrounding environment. (Vibrational relaxation occurs much more rapidly than does electronic relaxation.) Similarly, a molecule in an excited state can be converted to a state of the same multiplicity but of lower electronic energy level. This process is termed internal conversion. Again, the energy lost by the electron in the process appears as rotational and vibrational energy in the molecule. Where there are adequate numbers of excited state molecule in which this pathway is followed, the process may be demonstrated by detection of heat.

In addition to radiative decay (fluorescence or phosphorescence emission), intersystem crossing, or internal conversion, excited-state molecules can rid themselves of their excess energy by transferring it to other molecules or by undergoing chemical reactions. These processes will be discussed below.

The above radiative and radiationless pathways are not necessarily mutually exclusive. It is obvious, for example, in the case of phosphorescence, that intersystem crossing from the excited singlet state, first generated by photon absorption, to an excited-triplet state must occur before phosphorescence emission can take place. In addition, internal conversion from one excited singlet state to a lower-energy excited-singlet state can take place prior to fluorescence emission. Figure 1 shows a schematic representation of the possible pathways by which an excited state molecule can return to the ground state. As noted earlier, the particular pathway followed for a given excited state molecule depends on both the nature of the compound itself and environmental factors.

Energy Transfers

In addition to light emission, internal conversion, and intersystem crossing, an excited-state molecule can return to the ground state by transferring its excess energy to another molecule. This is one mechanism by which photosensitized reactions, which are very important in photobiological process, may occur. In the energy-transfer process, the acceptor molecule is promoted to an excited state. In biological systems, a one-step non-radiative transfer is the most probable process; however, a two-step radiative excitation transfer, in which the excited donor molecule emits a photon that is absorbed by the acceptor molecule, does occur in some bioluminescence phenomena. In the nonradiative process, the demotion of the excited donor molecule to a lower-energy electronic state, and the promotion of the acceptor molecule to a higher-electronic state occur simultaneously. In fluid solutions, the most common transfer processes are the so-called singlet–singlet and triplet–triplet transfers. In the singlet–singlet transfer, initially the donor molecule is in the first excited singlet state (S_1) and the acceptor molecule is in the ground singlet state (S_0). Following the transfer, the donor molecule is in the ground singlet state and the acceptor is in the first excited-singlet state. If "A" is the donor molecule and "B" the acceptor molecule, the transfer may be described thus:

$$A(S_1) + B(S_0) \rightarrow A(S_0) + B(S_1)$$

The triplet–triplet type transfer, in which the donor molecule is initially in the first excited triplet state (T_1) may be described by the following equation:

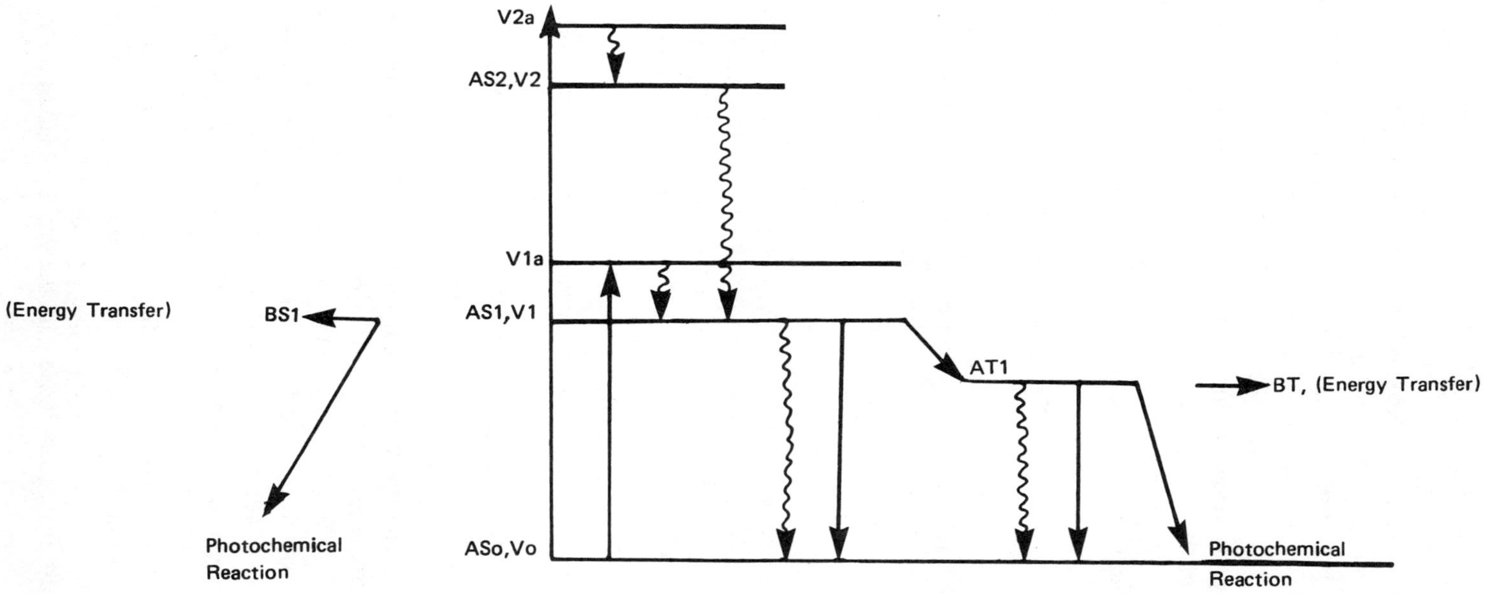

Figure 1. Simplified schematic depiction of the relaxation pathways available to excited state molecules. A molecule, *A*, absorbing radiant energy (↑) is elevated to an excited singlet state (e.g., S1, S2) with some vibrational energy level (e.g., V2a, V1a). Vibrational relation (VR) rapidly brings the molecule to the lowest vibrational level for the given excited electron state. Radiation-less decay can bring the molecule to a lower energy electron state by internal conversion or intersystem crossing to an excited triplet state. Alternatively, an excited singlet state molecule can return to the ground state by emission of radiation [fluorescence (↓)], by singlet–singlet energy transfer to another molecule (e.g., BS1), or by participation in a photochemical reaction. Similarly, the excited triplet can return to the ground state by intersystem crossing, emission of radiation [phosphorescence (↓)], triplet–triplet energy transfer to another molecule (e.g., BT1), or by participating in a photochemical reaction.

$$A(T_1) + B(S_0) \rightarrow A(S_0) + B(T_1)$$

Singlet–singlet transfers can occur over distances up to 100 Å. However, because the lifetimes of excited singlet states are so short, usually less than 10^{-7} sec, the concentration of acceptor molecules present must be on the order of 10^{-1} to 10^{-3} M for efficient transfer to occur. For triplet–triplet transfers to occur, a much closer proximity of donor and acceptor molecules is required. In deaerated solutions, however, the lifetimes of triplet states are usually quite long (10^{-6} to 10^{-3} sec). Thus, efficient triplet–triplet transfer in such systems can occur with concentrations of acceptor molecules much lower than those required for singlet–singlet transfer.

Direct Photochemical Reactions

Photobiology is concerned with responses of biological systems to irradiation with light. Since biological responses are dependent upon the occurrence of chemical reactions, it is reasonable to assume that responses of biological systems to light are based upon chemical reactions that take place as a consequence of the absorption of light by molecules in the system. As we have seen, absorption of light by a molecule results in the molecule being in an excited state. The above radiative and radiationless electronic relaxation pathways result in the return of an excited molecule to its original ground state. Thus, except for energy transfer, which results in the continued presence of excited molecules in the system, with these processes no chemical changes take place and no net biological response to light exposure would be expected. As noted earlier, other mechanisms are available for relaxation of excited-state molecules which involve bond breakage or bond formation. Such photochemical processes do result in alterations of molecules and are, therefore, the means by which photobiological responses are produced.

In photobiology, photochemical reactions are commonly subcategorized as either direct (i.e., the original light-absorbing molecule undergoes a chemical change) or photosensitized (the original absorbing molecule causes a chemical change to occur in other molecules in the system). In application, this distinction may be hard to make, since technically some "direct" photochemical reactions may, by definition, also be photosensitized reactions. For example, if the originally excited molecule $A(exc)$ reacts with another ground state molecule in the system $B(gs)$ to form a covalent bond: $A(exc) + B(gs) \rightarrow AB$, molecule B is clearly changed, but so is molecule A. To get around this difficulty, in photobiological contexts, photosensitized reactions are usually defined in terms of reactions occurring because of the presence in the system of an agent not normally endogenous to the system (e.g., a drug or other exogenously added substance). Photosensitization will be considered in greater detail below.

A molecule in an excited state has different chemical and physical properties from the same molecule in the ground state. Thus, the types of chemical reactions engaged in, the reactive intermediates, and the products produced with an excited state molecule are usually quite different from those with the ground state of the molecule.

Although the initial chemical reaction occurring with an excited state molecule sometimes results in immediate formation of a stable product, photochemical reactions usually proceed via the production of transient reactive intermediates, which then react further to produce a final product. In this latter context, the factors influencing formation of the initial photoproduct (e.g., ions, radicals, and other unstable species) and those influencing later reactions occurring with the intermediate can be quite different.

The types of initial or primary photochemical reactions that take place with excited state molecules are limited. Among the more commonly occurring primary reactions are electron transfer, hydrogen abstraction, cis-trans isomerization and other types of bond rearrangement, bond cleavage, and cycloaddition. Examples of some of these are shown in Figure 2.

As noted earlier, in some cases the initial photochemical reaction results directly in a stable product, while in others intermediates are produced which go on to participate in additional chemical reactions. The additional reactions that may take place (often called dark reactions, since no further light exposure is required for them to occur) are numerous and depend upon the molecular composition of the system and other factors (temperature, pH, etc.).

Photosensitization

Photosensitized reactions are those in which light absorption by a molecule results in chemical modification of another molecule in the system. In photobiology, the term "photosensitizer" (i.e., the light-absorbing agent causing changes in other molecules) is usually reserved for agents not native or endogenous to the system in question (i.e., the reaction observed upon exposure to light in the presence of the exogenously added agent does not occur in its absence).

Hundreds of chemical compounds are capable of acting as photosensitizing agents.[1] Among the compounds demonstrated to have photosensitizing properties are a wide variety of drugs used in routine medical practice, as well as in industry, agriculture, and the home. The majority of biologically active photosensitizers contain a tricyclic ring; they are highly conjugated molecules. The relatively long life of the triplet state of a molecule, compared to that of the singlet state, enhances the potential for interactions with other molecules. Thus, another characteristic of most biologically active photosensitizers is a high degree of efficiency of intersystem crossing from the excited singlet to the triplet state.

In biological systems there are essentially two major classes of reactions in which photosensitizer triplets participate.[2] In one, the sensitizer interacts directly with another molecule resulting in transfer of either a hydrogen atom or an electron. These are the so-called Type I radical reactions. The other class of reactions, the so-called Type II reactions, involve interaction of the sensitizer triple with molecular oxygen. The most frequently occurring of these interactions is transfer of energy from the excited sensitizer molecule to the oxygen molecule, producing singlet-state oxygen. This is

e⁻Transfer

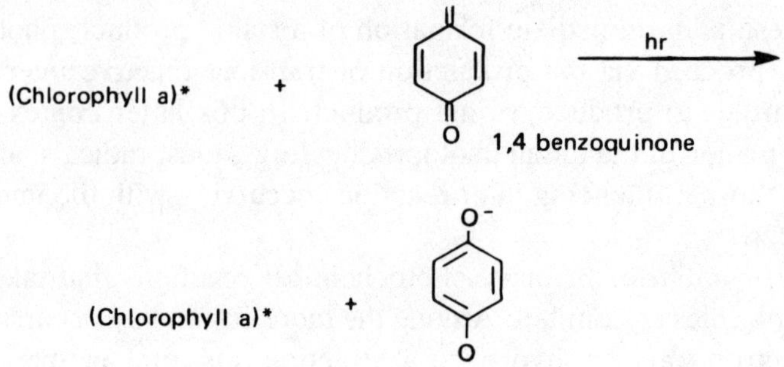

H Abstraction

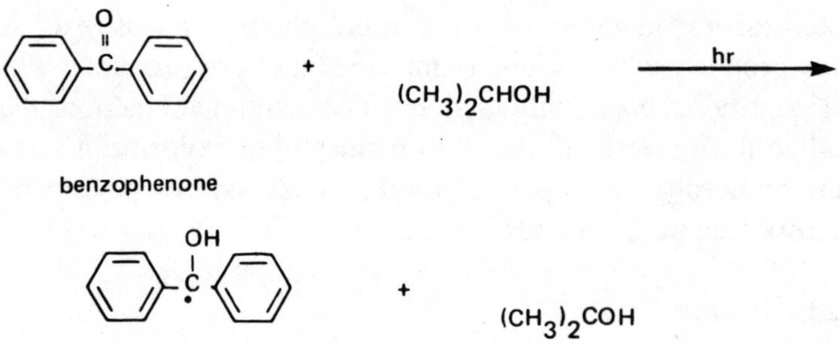

Cis-Trans Isomerization

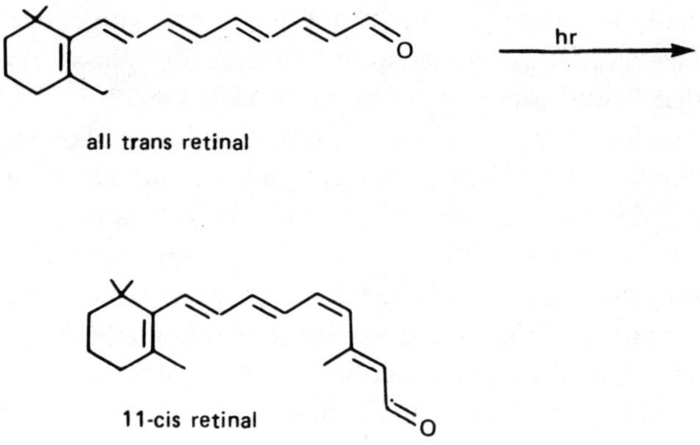

Bond Cleavage

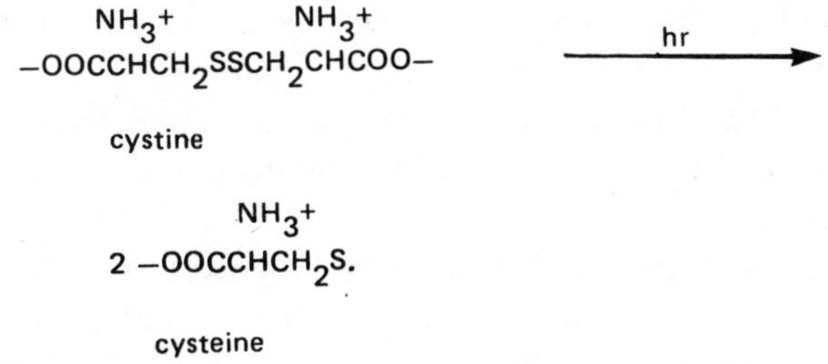

Cycloaddition

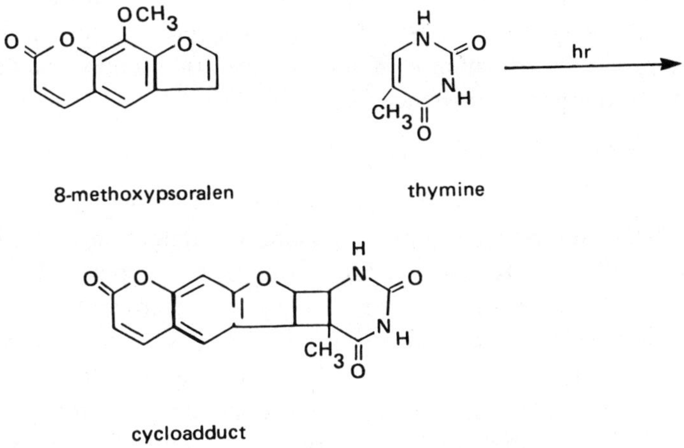

Figure 2. Examples of primary photochemical reactions: e⁻ transfer; H abstraction; cis-trans isomerization; bond cleavage; and cycloaddition.

a very rapid process and accounts for most of the quenching of excited triplets by oxygen; however, a less-efficient process, involving electron transfer to the oxygen, producing the superoxide ion, can also occur. (It should be noted that sensitizer energy transfer to molecules other than oxygen can occur and can account for responses of biological systems to the presence of photosensitizers. The relatively small size and ready diffusibility of the oxygen molecule, however, tend to result in it being the more likely acceptor in terms of being able to get in appropriate proximity to the excited sensitizer donor.)

Free radicals generated by reactions with triplet sensitizer molecules can participate in a series of chain reactions that can lead to severe cell damage. Moreover, singlet oxygen generated by photosensitizers is relatively long-lived, extremely reactive, and capable of reacting with many substrates, thus propagating a broad spectrum of chemical reactions.

In man, the most common photosensitizing reactions involve the skin, although ocular tissues may also be affected. Phototoxic reactions may also affect internal organs. Photoallergic reactions, which are delayed responses, also involve photosensitized processes in which formation of specific compounds that react with proteins to form antigenic substances takes place.

Measures of Photobiological Response

Analysis of the mechanisms involved in a photobiological response can be quite complex. The time between light exposure and detection of a response can range from a few milliseconds to years. Furthermore, the response detected is usually the end product of a series of molecular events. While absorption of light is necessary for a photochemical event to occur, it does not ensure that the required event will take place. As noted above, photochemical reactions are not the only means by which excited-state molecules can return to the ground state. Moreover, even if a primary photochemical reaction that is capable of initiating the chain of events leading to the ultimate biological response does take place, in most cases a number of alternative pathways proceeding from the initial reaction are available as well. Finally, even if light absorption does lead to generation of a product required to produce a response, the amount of this product produced may not be adequate for the response to be detected. Thus, the events leading from absorption of light in a biological system to the appearance of a detectable response are dependent upon the numbers of photons absorbed, the numbers of molecules that undergo the particular photochemical reactions leading to the production of the response, and the numbers of altered molecules required to effect the response.

In many cases, the effects of exposure of biological systems to light involve inactivation process that take place with whole cells, organelles, particular biological functions, or particular biological molecules. When the occurrence of such events can be quanitified, for example, by determining numbers of cells surviving or measuring extent of enzyme inactivation following light exposure, some of the parameters influencing production of the effect can be evaluated.

One analytical technique commonly applied is to determine the action spectrum for the event of interest. The action spectrum is a plot of the reciprocal of the number of incident photons required to produce a given effect compared to the wavelength of the radiation employed. The number of incident photons can be derived from radiometric measurements of the intensity of the radiation impinging upon the system. Radiometric measurements of intensity (or irradiance) describe the radiation incident on a surface normal to the beam of light in terms of the power applied (in watts) per unit area (cm^2):

$$\text{Intensity} = \frac{\text{Power}}{\text{area}} = \frac{\text{Watts}}{\text{cm}^2}$$

Since the power is the amount of energy in joules (J) delivered per unit time (in seconds),

$$\text{Intensity} = \frac{\text{J}}{\text{sec} - \text{cm}^2}$$

As noted earlier, the energy of a given photon is related to its wavelength by the equation:

$$E = \frac{hc}{\lambda}$$

Thus, if the wavelength and intensity of the radiation impinging upon a system are known, the number of photons incident upon the system can be calculated. The number of events of interest that occur, divided by the number of incident photons, is the "photosensitivity" (σ) of the system:

$$(\sigma) = \frac{\text{number of events}}{\text{number of incident photons}}$$

Thus, an action spectrum for a given response is a plot of the photosensitivity of the system against wavelength. In an action spectrum, peaks represent the wavelengths most efficient for producing the effect, that is, wavelengths that require the fewest photons to get the job done. The action spectrum for an effect frequently resembles the absorption spectrum of the chromophore that absorbs the radiation and is responsible for the effect. However, differences between the action spectrum and the absorption spectrum of the chromophore responsible for the effect can arise if, for example, the absorption characteristics of the chromophore *in vivo* are different from its absorption characteristics *in vitro*, or if significant absorption of wavelengths of interest occurs with other components in the system. Thus, while matching of an action spectrum for an event with the absorption spectrum of a chromophore in the system is suggestive of a causal relationship, it is by no means proof of such a relationship.

If, instead of the number of events occurring per incident photon, one examines the numbers of events that occur divided by the photons *absorbed*, this is the "quantum yield" (Φ) for a given effect:

$$\Phi = \frac{\text{number of events of interest}}{\text{number of photons absorbed}}$$

On the molecular level, the Starck–Einstein Law of photochemical equivalence tells us that one photon can, at most, cause one molecule to undergo a given photochemical process. However, since a number of processes are available to excited-state molecules, absorption of a photon does not necessarily lead to the particular event of interest. Thus, the quantum yield for a given process will always be ≤ 1. The number of photons absorbed by a system can be derived from the absorption cross-section for the systems, where:

$$s = \frac{\text{the number of photons absorbed}}{\text{the number of incident photons}} = 3.82 \times 10^{-21}\ \epsilon$$

(ϵ is the molar extinction coefficient empirically derived as described earlier.) The absorption cross-section, photosensitivity, and quantum yield for a given effect are clearly related to one another. This relationship can be expressed by the equation:

$$\sigma = s\ \Phi$$

As noted above, a number of factors can influence the efficiency with which a light-initiated biological response occurs (e.g., the number of photons absorbed, the number of molecules that undergo the appropriate photochemical reaction, and the degree of response per altered molecule). "Response efficiency" is a term frequently encountered in photobiology. It is similar to quantum yield and is expressed as the observed response divided by the number of incident photons.

All of the above measures relate photobiological response to the *numbers* of photons either incident on, or absorbed by, the system. These observed relationships between numbers of photons, rather than the rapidity with which they are delivered, are the basis of the so-called Bunsen–Roscoe reciprocity relationship. Since many photobiological responses are linearly related to the numbers and kinds of photochemical reactions, this relationship has come to be accepted as a basic principle in photobiology; however, it should be noted that reciprocity may not hold in some complex responses.

Photobiology of Mammalian Cells

The effects of nonionizing radiation on a cell depend upon the specific chemical composition within the cell, that is, on the presence or absence of molecules capable of absorbing the radiant energy. If cells do not contain compounds that absorb at a given wavelength, they will transmit that wavelength. The nucleic acids and most proteins are essentially transparent to, and thus transmit, visible light. They can, however, absorb certain wavelengths in the UV region of the electromagnetic spectrum. In DNA, the purine and pyrimidine rings of the nucleic acids absorb light maximally in the 245–280-nm range, conferring an absorption maximum on DNA itself of around 260 nm. Absorption of UVB and UVC radiation may result in formation of dimers between adjacent thymine residues on a DNA strand or in covalent cross-linking of proteins to DNA molecules. Absorption of UVA by exogenous photosensitizers can also result in DNA alterations. Light-induced alterations in DNA can result in initiation of enzymatic repair processes or may lead to cell mutations or even death, depending upon the nature and extent of damage.

Protein molecules are also capable of undergoing alterations as a consequence of ultraviolet light exposure. These can include oxidation of certain amino acids and covalent cross-linking, which may result in loss of specific protein function, (e.g., enzyme activity) aggregation, or conformational alterations. As with DNA, the alterations produced can arise from direct absorption of photons by chromophores native to the molecule (i.e., aromatic amino acids such as tryptophan and tyrosine) or from interactions resulting from light absorption by an exogenous photosensitizer.

Other molecules that may be present in cells, such as rhodopsin and hemoglobin, have absorption maxima in the visible region and can thus be damaged by visible radiation at their specific absorption wavelengths.

It should be reiterated that damage to cells arising as a consequence of absorption of light by either a native or exogenous chromophore may not be limited to the absorbing species itself. As noted in preceding sections, the absorbed energy may be transferred to other molecules, which may then engage in photochemical reactions, free radicals may be generated, and a variety of dark reaction pathways may be initiated that can ultimately result in damage to structures, systems, and components of cells which of themselves would not be directly affected by radiation exposure. Thus, for example, membrane lipids may be oxidized, resulting in alterations in permeability and/or in transport function, enzymes may be inactivated, severely impairing metabolic function, DNA alterations can lead to impaired or improper transcription, and damage to RNA may affect translation processes.

Photobiology of the Skin

The skin has certain built-in protective devices to minimize damage that might occur due to light exposure. The stratum corneum reflects up to 5–10% of incident solar radiation. In addition, it is composed principally of a fibrous protein, keratin, which not only absorbs significantly in the UVB and UVC regions of the spectrum, but also, because of its structural characteristics, scatters most visible radiation. Keratinocytes, because of their native chromophores (most notably DNA), serve to absorb incident ultraviolet radiation further. As a consequence, only 5–10% of solar radiation in the 290–310-nm wavelength range actually penetrates to reach the basal cell layer and superficial dermal vasculature. On the other hand, significant transmission through the epidermis of ultraviolet radiation above 330 nm does occur. In Caucasian skin as much as 50% of incident UVA radiation may be transmitted to the basal cell layer and dermis. In blacks, this transmission is significantly reduced, so that in persons with very dark skin, as little as 5–10% of incident UVA may penetrate the epidermis.

The presence of melanin in the skin serves to protect this organ further from light-induced damage. This pigment, which is synthesized in specialized epidermal cells (melanocytes) and packaged in organelles called melanosomes, is transported by dendritic processes of the cell to keratinocytes and is then "injected" into the keratinocytes.[3] Melanin clearly plays a role in protection against the effects of sunburn radiation (UVB). It may function in this regard in several ways. The large size and opaque nature of the pigment molecules may serve to block and scatter UV radiation physically.[4] Melanin may also play a role in trapping free radicals generated as a result of UV exposure and may further serve as a sort of *in vivo* thermopile, absorbing UV radiation and converting it to heat.[4]

Increases in epidermal melanin pigmentation (tanning) occur upon exposure to UV radiation. Two different types of tanning can be distinguished, both on the basis of when they occur with respect to onset of light exposure and on the mechanisms operative in their production. Immediate tanning, which is most effectively induced by UVA radiation, but can also be induced by visible radiation,[5] begins very shortly after onset of radiation and peaks within 1–2 hr.[6] The duration of immediate tanning is dependent upon the length of light exposure, fading within a matter of minutes with short exposure and lasting in excess of 36 hr with protracted exposure to sunlight or exposure to high-intensity long-wave UV radiation sources.[7] Immediate tanning evidentally does not involve new melanin production, but apparently arises from alterations in melanin already present in the epidermis. Spectroscopic studies indicate that these alterations most likely involve reactions in which parts of melanin molecules are oxidized to yield unstable semiquinone-like free radicals.[8] In addition, in human skin, immediate tanning is accompanied by movements of melanosomes from perinuclear loci in melanocytes into the dendritic processes and increases in the transfer of the melanosomes into keratinocytes.[7]

The ability to exhibit immediate tanning is a function of both the unexposed inherited coloration of the individual (i.e., constitutive skin color) and the extent of tanning previously acquired. Skin previously exposed to radiation (i.e., already tanned to some extent) shows a more striking immediate tanning response than does unexposed skin. Similarly, the darker the normal color of the skin, the more impressive is the immediate tanning demonstrated.

The other type of tanning, delayed tanning, becomes noticeable much later after radiation exposure than does immediate tanning, reaches a peak much later, and lasts much longer. It first becomes apparent between 48 and 72 hr after UV exposure, peaks at 13 to 21 days, and subsides over a period of several months.[6] In contrast to immediate tanning, which is caused by UVA radiation but apparently not by UVB, the major peak in the action spectrum for delayed tanning resides in the UVB region.[7,9] However, UVC, UVA, and visible light in the blue region are also capable of inducing delayed tanning.[7,9] As with immediate tanning, the relative doses of radiation required to produce delayed tanning depend on both the constitutive nature of the exposed skin and its history of previous exposure. Delayed tanning is a result of new melanin production, which is associated with the following:

1. An increase in the size of melanocytes and increased growth and branching of dendritic processes
2. An increase in melanosome production within melanocytes
3. An increase in the size and melanization of melanosomes
4. An increase in tyrosinase activity in melanocytes
5. An increase in the rate of transfer of melansomes to keratinocytes, resulting in a greater number of melanin granules in the epidermis.[10]

(Delayed tanning is also associated with increases in number of melanocytes; however, in man, multiple exposures to UVB radiation are required to demonstrate this.[11])

Although, as noted above, UVA radiation can induce delayed tanning, there are differences between UVB and UVA responses. Larger doses of UVA than of UVB are required to produce the effect.[7, 9] In some individuals, slightly less UVA radiation may be required to induce tanning than is required to produce an erythemal response, while with UVB erythemogenic doses are required.[7, 9] An additional difference between skin that has been tanned by repeated exposure to UVA radiation and skin tanned by UVB relates to increases in skin thickness that are considerably more prominent with UVB-induced tanning.[12, 13]

Delayed tanning serves a photoprotective function, the increase in melanin present in the superficial skin layers serving to decrease transmission of ultraviolet radiation to the deeper layers. This protective effect is further enhanced when thickening of the stratum corneum and epidermis accompanies melanization.[9, 12, 13] This latter response results from a sustained increase in epidermal mitosis associated with increases in DNA, RNA, and protein synthesis that follow a single exposure to ultraviolet radiation. Prior to these increases in mitotic and synthetic activity, a transient decrease in both is observed. Most studies on these phenomena have employed single doses of UVB radiation in either human or animal skin.[14, 17] However, similar thickening of the epidermis and stratum corneum has also been seen with exposure to UVC and UVA radiation.[18] The pathway leading to epidermal proliferation is not known; however, UV radiation has been shown to induce increases in ornithine decarboxylase,[19, 20] which is the rate-limiting enzyme in polyamine biosynthesis[21] and may be a factor in promoting DNA synthesis.[21, 22]

Despite its built-in protective mechanisms, the skin is still susceptible to a whole host of disorders in which light exposure may play an etiologic and/or contributory role. Ultraviolet-induced erythema or sunburn is the most ubiquitous of the photobiological responses evidenced by skin. Virtually all humans are susceptible to this inflammatory response to a greater or lesser degree, depending on the extent of skin pigmentation.

Ultraviolet-radiation-induced erythema is perhaps one of the most extensively studied photobiological responses in humans. While a degree of transient redness may be seen within minutes of UV exposure, the major erythemal response is delayed in onset, usually not appearing until several hours after exposure. The nature and exact location of the photochemical event initiating the erythemal response are unknown.

One theory suggests that the event occurs in the epidermis, where light absorption by keratinocytes may result in generation of as yet unidentified vasoactive mediators, which then diffuse to dermal blood vessels and produce vasodilation. This "diffusion" hypothesis is supported by the observation that much of the radiation of wavelengths capable of eliciting an erythemal response is absorbed by the epidermis. Moreover, there is a latent period between exposure and development of erythema which would also be consistent with the required diffusion of active agents from one site to another. Nevertheless, direct injury of vascular tissue or to other dermal sites cannot be excluded, and in some cases, such as with UVC-induced erythema, may actually be the main mechanism operating.[7, 23]

All wavelengths of ultraviolet radiation are capable of inducing skin erythema; however, their relative effectiveness in this regard is highly variable, as can be seen by examining the minimal doses required to produce skin erythema (MED). For wavelengths in the UVC range (200–290 nm) the MED is on the order of 10^{-1} to 10 mJ/cm_2, while for wavelengths in the UVB (290–320 nm) and UVA (320–400 nm) ranges, the MEDs are on the order of 10–100 mJ/cm^2 and 10^4–10^5 mJ/cm^2, respectively.[7, 25]

Erythema produced by UVC radiation begins within 4–6 hr of exposure and fades within 12–36 hr. It is usually not associated with increases in skin pigmentation, but may result in desquamation. No UVC from the sun reaches the earth's surface. Thus, where UVC-induced erythema occurs, it is derived from either intentional or accidental exposure to man-made sources of these wavelengths, most notably the cold quartz "germicidal" lamp.

UVB radiation is present in solar radiation reaching the earth's surface. UVB-induced erythema usually becomes evident within 2–6 hr after exposure, and may last up to 120 hr. In contrast to UVC-induced erythema, UVB erythema is usually associated with increased skin pigmentation.

UVA radiation can also induce erythema, but, as noted above, as much as 1,000–10,000-fold greater dose is required than with UVB radiation.

As previously noted, the primary photochemical event initiating the UV-induced erythemal response remains unknown; however, a number of events associated with the response have been described. The vascular response to ultraviolet radiation is biphasic in experimental animals, with a transient immediate increase in vascular permeability (and, sometimes, vasodilation) followed by a prolonged increase in both dilation and permeability some 2–8 hr later. The early phase of dilation may be related to release of histamine, since pretreatment with antihistamines can reduce this response.[26, 27] Serotonin may also play a role in the initial response.[28, 29] The delayed response, however, is not suppressed by inhibitors of any of these agents, nor is it suppressed by inhibitors of proteases or plasminogen activators.[7, 25] On the other hand, both topical and intradermal applications of the prostoglandin synthesis inhibitor, indomethacin, have been shown to both delay and decrease the intensity of UVB induced erythema, even in man. UVC-induced erythema was also affected by indomethecin treatment, although not to the extent that UVB-induced erythema was.[30, 31] UVA-induced erythema was not affected at all.[32] Thus, prostaglandins are evidentally involved to some extent in mediation of UVB and UVC erythema, but appear not to be involved in UVA erythema.

Prior to onset of the delayed vascular response, a cellular infiltrate consisting mainly of polymorphonuclear leukocytes can be found in UV-exposed skin. It has been suggested that these cells may release agents involved in development of the latent response. Animals made neutropenic prior to UV exposure, however, still evidence the delayed response, making the role this infiltration plays unclear. Still another possible source of substances that cause the vascular responses may be lyososomal enzymes or other agents released from keratinocytes damaged by UV exposure.[33, 34]

Chronic ultraviolet exposure has been correlated with both premature aging and development of skin cancers. Changes referred to as actinic damage or premature aging include wrinkling, yellowing, increased skin fragility, and telangiectasia. Caucasians are more susceptible to these types of damage than are blacks. While melanin-producing ability is one of the major factors in determining susceptibility to actinic damage, other factors, including repair function, may also be involved. Changes associated with premature aging involve the epidermis, dermis, and the structures contained therein. In the dermis, sunlight induces a decrease in soluble collagen, with a concomitant increase in elastin. In addition, chronic sunlight exposure produces changes in the elastin fibers, including fragmentation, curling, hyperplasia, hypertrophy, and, finally, massive degeneration. These changes, termed solar elastosis, occur only in sun-exposed skin, primarily in the papillary dermis. Experimental induction of solar elastosis in humans remains to be accomplished, but changes of a similar nature have been induced in mice and rats.[35-37] Whether UVB or UVA or both are responsible for development of solar elastosis remains to be determined. One hypothesis as to how UV exposure may lead to premature aging has to do with the ability of UV to produce cross-linking in DNA. In a general theory of aging, it has been postulated that cross-linking reactions are responsible for age-dependent changes. These cross-linking processes may involve a wide variety of agents in cells and a gradual accumulation of cross-linked material may take place. With DNA, cross-links within and between DNA molecules, and between DNA and protein molecules, are continuously occurring and being repaired. However, repair processes are not always complete. Thus, aging may be related to accumulation of DNA adducts. Since UV exposure can produce cross-linking of DNA, processes involved in "premature" aging may simply be an acceleration of those occurring with "natural" aging.

The relationship between ultraviolet radiation exposure and the development of skin cancer has long been appreciated. Numerous reviews have appeared concerning work in this area, including one in the present volume, and an in-depth discussion will not appear here. Evidence shows that ultraviolet radiation exposure may be an etiologic and/or contributory factor in the development of all three of the major skin cancers in humans (basal cell and squamous cell carcinomas and malignant melanomas). Studies in animal systems further suggest that skin photocarcinogenesis is influenced by the wavelengths of the UV radiation, total dose applied, the mode of its delivery, and other factors including genetic, immunologic, and environmental influences.[38-47]

Aside from sunburn and photocarcinogenesis, a whole host of skin disorders may be initiated or aggravated by light exposure. Some of these disorders require light for their expression, for example, phototoxic and photoallergic responses. In others, light exposure may exacerbate a skin condition already present but derived from another source. In still others, light exposure may precipitate a skin reaction that is only one of a constellation of phenomena associated with some systemic disorder.

Adverse skin responses that occur because of the interaction of light with some agent not normally present in the tissue (e.g., a systemically administered drug or some agent coming in contact with the skin externally) are frequently grouped together and called drug-induced photosensitivity. Implicit in the definition of such responses is that neither the foreign agent nor the light exposure by itself is capable of eliciting the response. Not all cases of drug-induced photosensitivity are the result of the foreign agent acting directly as a photosensitizer. In some cases, the foreign agent may induce production of photosensitizers by the host (e.g., porphyrin biosynthesis) or it may induce pathologic changes that potentiate skin photosensitivity (e.g., drug-induced lupus erythematosus or drug-induced pellagra).

Among skin photosensitivity responses in which the foreign agent acts as a photosensitizer, two broad categories of reactions can be differentiated—phototoxic responses and photoallergic responses. Clinically, it may be difficult to distinguish between the two phenomena; however, differences do become evident under ideal circumstances. Since photoallergic reactions involve an immunologic response, for these reactions to become manifest the individual exhibiting the response must have been exposed to the causative agent at least once prior to the exposure that produced the reaction. Moreover, as with any immunologic response, there must have been an incubation period between the initial exposure and the exposure that led to the reaction. In contrast, phototoxic reactions can occur upon first exposure to the causative agent and require no incubation period. Individuals exhibiting photoallergic reactions to a given agent frequently display a similar reactivity to other agents that are structurally related to it. This is not the case with phototoxic reactions. The concentrations of agent required to elicit a photoallergic reaction are usually quite small compared to concentrations required to produce a phototoxic reaction. In addition, while all agents that elicit a photoallergic response undergo some structural alteration to do so, this is usually not the case with agents causing phototoxic reactions. In theory, an agent capable of eliciting a phototoxic reaction should cause such a response in 100% of the population, while the occurrence of a photoallergic reaction to an agent will be sporadic and highly dependent upon the nature of the host.

Photoallergic reactions are cell-mediated delayed hypersensitivity responses. Evidence from animal systems suggests that the chemical in the skin absorbs light and is converted to a photoproduct which then binds to either soluble or membrane bound protein to produce an antigen.[48-50] Halogenated salicylanilides are the most common group of chemicals known to elicit a photoallergic response. In the past, these agents were frequently added to an assortment of household and cosmetic products because of their antimicrobial properties. Examples of these types of chemicals that may cause human photoallergic reactions include Fentichlor, Multifungin, 6-methylcoumarin, and 3,5,3′,4′-tetrachloro-salicylanilide (by far the best-studied).[51-54] Another agent that evidentally can cause photoallergic reactions is musk ambrette found in fragrances, particularly aftershave lotions.[55,56]

Phototoxic agents may be classified into two categories, based on whether or not oxygen is required for their activity. Photodynamic agents require oxygen to elicit a phototoxic reaction while non-photodynamic agents do not. Phototoxicity reactions may also be categorized according to the subcellular locus affected. Principally DNA, and both lyosomal and plasma membranes, are the sites of action of phototoxic agents. 8-methoxypsoralen, for example intercalates between DNA base pairs, and then, upon light exposure, photochemically forms either monoadducts or cross-links by cycloaddition to the 5,6 position of the pyrimidine ring.[57,58] After such addition, DNA synthesis is inhibited and repair processes are activated.[58]

Membrane-associated phototoxic action can often be demonstrated by using assays for red blood cell hemolysis following light exposure. An example of an agent that, in the presence of light, produces oxidative membrane damage is protoporphyrin.[59] Among agents that can produce skin phototoxic reactions are certain dyes (e.g., eosin and anthraquinone), coal tar derivatives such as phenanthranene and acridine, furocoumarins (e.g., 8-methoxypsoralen), and drugs such as tetracyclines, sulfonamides, and phenothiazines. More detailed accounts of our knowledge of phototoxic and photoallergic responses can be found in other articles in this volume.

Photochemistry and the Eye

The First Law of Photochemistry states that light must be absorbed before photochemistry can occur. Thus, whether or not nonionizing radiation will affect a given cell or tissue depends upon the presence or absence of chromophores capable of absorbing the radiant energy. Other than the skin, the eye is the only organ or tissue in the body particularly sensitive to the wavelengths of nonionizing radiation (> 280 nm) normally present in the environment. The normal cornea, aqueous humor, ocular lens, and vitreous are almost completely transparent to all but the shorter wavelengths of visible light. (The aging lens does, however, absorb increasing amounts of visible radiation in the shorter wavelengths.) Photochemical damage to these ocular components from visible radiation, therefore, would not be anticipated. On the other hand, the retina, with its photoreceptor rods and cones and macular pigment, is abundantly provided with chromophores whose function is to absorb visible light. As a consequence, this tissue is susceptible to both direct and photosensitized damage from visible radiation.

While very few visible-wavelength-absorbing chromophores are present in ocular tissues anterior to the retina, these tissues, particularly the ocular lens, do have the ability to absorb varying amounts of ultraviolet radiation. With aging, the ocular lens progressively develops chromophores capable of absorbing longer-wavelength UV radiation, and thus, in the mature individual, provides an effective filter to protect the retina from these wavelengths.[60, 61] The lens, overall, probably sustains the greatest amount of photochemical change of all ocular tissues during a lifetime of exposure to ambient UV radiation. In the very young eye, where the lens has not yet become an effective UV filter, and in the aphakic or pseudophakic individual; the retina can be exposed to, and potentially damaged by, ultraviolet light.

Normally, ambient infrared radiation (750 nm–1 M) by itself is incapable of causing damage to ocular tissues. The only possible effects on ocular tissues of infrared radiation at ambient levels would be those associated with small temperature elevations in absorbing tissues (mainly the retina and lens) which may serve to enhance UV-induced photochemical damage. On the other hand, very high exposure levels (e.g., from CO_2 or Nd:YAG lasers) can produce significant damage, due to both generation of shock waves (acoustic gradients) and intense heating.[62] Exposure to high levels of infrared radiation from blast furnaces has also been shown to cause cataract development (Glass Blower's Cataract).

As with the skin, photoinduced changes occurring in ocular tissues can be attributed to two mechanisms: (1) a direct or intrinsic process wherein radiation is absorbed by chromophores native to these tissues (e.g., nucleic acids or aromatic amino acid residues), and (2) an indirect or photosensitized process wherein radiation is initially absorbed by an extrinsic compound or drug that has entered one or more of the ocular tissues.

Numerous drugs have been suspected of having as side effects the potential for causing photosensitized ocular damage.[63] However, only two groups of compounds—the phenothiazines and the psoralens—have been clearly demonstrated to be intraocular photosensitizing agents, capable of causing photochemical damage to the choroid, the retina, and the lens in humans, as well as in experimental animals.[64-81] The fact that relatively few drugs known to have photosensitizing properties in other systems have been demonstrated to have such effects in the eye may be related to screening effects of the blood–aqueous and blood–retinal barriers. For example, it is well-known that it is difficult, if not impossible, to obtain effective concentrations of certain antibacterial, antimycotic, and antibiotic drugs in the interior of the normal eye; this has been attributed to the relative impermeability of the above-mentioned barriers. In addition, it has been demonstrated that the cornea, and particularly the ocular lens as it ages, provide effective filters for UV radiation and even shorter wavelengths of the visible spectrum,[61] thereby preventing potential photosensitizing activities of drugs accumulating in the posterior half of the globe that may be "activated" by these wavelengths. On the other hand, since the lens does not shed any of its cells and the neural retina does not regenerate, any photosensitizing agent accumulating in these tissues might be a potential hazard, particularly if it becomes photobound to macromolecules present in the tissues.

An example of a drug that can be photobound in both these tissues is 8-methoxypsoralen (8-MOP), which has been used extensively for the treatment of psoriasis and vitiligo.[82, 83] This form of phototherapy, commonly referred to as psoralen–UVA, or PUVA, therapy, involves ingestion of 8-MOP (or related compounds) followed by exposure to UVA radiation (320–400 nm) for short periods of time. In animals (rat, dogfish, and monkey), 8-MOP can be found in a variety of ocular tissues within 2 hr after administration of a single dose equivalent to the human therapeutic level. The drug can also become photobound to DNA[71] and lens proteins if the animals are exposed to ambient roomlight levels of UVA radiation.[73, 74, 79–81] As noted above, the mature ocular lens is a very effective filter for UV radiation in most mammals, including man. Thus, in mature animals, with their lenses intact, no photobinding of 8-MOP occurs in the retina. In aphakic and pseudophakic animals, and in young eyes where the ocular lens still transmits significant UV radiation, however, UVA radiation can penetrate to the retina and 8-MOP photobinding can occur.[76–78, 81]

Cataract formation arising as a consequence of PUVA therapy has now been documented in humans as well as in experimental animals.[65–76, 79, 80] The lens proteins from the cataracts from these PUVA patients, upon being subjected to high-resolution phosphorescence spectroscopy, showed phosphorescence peaks identical in shape and lifetime to those previously reported for the 8-MOP lens protein photoproduct seen in experimental rat PUVA cataracts. It should be borne in mind that 8-MOP can be found in the lens for only 24 hr after ingestion, if the eye is protected from UVA radiation. Thus, the simple expedient of providing PUVA patients with proper UV filtering glasses and instructions to put them on as soon as they ingest the drug and wear them for at least 24 hr, removes the risk for cataract development otherwise associated with this therapeutic regimen. (The glasses must be worn indoors as well as out of doors, since there is sufficient UVA radiation in ordinary fluorescent lighting to photobind 8-MOP.[69, 79])

In preliminary studies on the effects of 8-MOP and UVA exposure in the retinas of albino and pigmented mice, several organelles were identified as subject to damage by such treatment.[84] These include the rod outer segments (ROS), the ellipsoid region of the rod inner segments, and the receptor nuclei. The damage observed included vesiculation and disruption of the ROS, swelling of the ellipsoid as a whole, with damage to the mitochondria, and necrosis of the nuclei. Exposure to UVA radiation alone selectively caused more damage to the ROS, while the combination of 8-MOP and clinical UVA exposure resulted in far greater necrosis of the nuclei, in addition to ellipsoid and ROS damage.[84, 85] Photobinding of 8-MOP to DNA and cross-linking produced upon irradiation could account for the observed necrotic process.

Compared with other photosensitizers, a great deal is known about the specific ocular effects of 8-MOP. Other drugs that may have photosensitizing properties in ocular tissues that may be relevant to human applications include Declomycin,[63] griseofulvin,[63] chlorpromazine (and other phenothiazines), nalidixic acid,[63, 86] allopurinol,[87, 88] trifluorperazine,[89] and the anthracycline antibiotic, adriamycin, widely used in cancer chemotherapy.[90, 91] For a further discussion of ocular photobiology the reader is referred to the specific article on that topic in this volume.

Phototherapy and Photochemotherapy

No overview of photobiology, with particular emphasis on subjects of clinical relevance, would be complete without some reference to the use of light in therapeutic modalities. However, since specific articles in this issue consider this topic in depth, we will discuss them only briefly here.

Among the most ancient uses of light in conjunction with a drug for treatment of a pathologic condition has been mentioned already: psoralens and sunlight for the treatment of certain skin disorders.

Psoralens, which belong to a class of compounds called furocoumarins, may be derived from a variety of natural sources (e.g., celery, cloves, parsnips, and figs).[92] A number of psoralen derivatives have now also been synthesized in the laboratory.[93, 94] Psoralen ingestion or topical exposure, followed by UVA, produces erythema in skin of human and experimental animals and has been reported to account for some skin phototoxic reactions.

In skin, the psoralens used in photochemotherapy appear to exert part of their effects by intercalculating in DNA and then, upon light exposure, forming covalent cyclobutane-type adducts with DNA bases.[57, 58] In vitiligo, PUVA therapy induces melanocyte stimulation, resulting in repigmentation of affected skin areas. In psoriasis, on the other hand, the affected epidermis replicates more rapidly than normal, PUVA treatment (and some other therapeutic modalities such as the Goeckerman coal tar regimen) serves to inhibit DNA synthesis in drug-treated and UVA-exposed sites. Among the drawbacks associated with PUVA therapy are a possibly enhanced risk for skin cancer[95] and potential damage to ocular tissues, as discussed above.

Other types of photochemotherapy currently used include topically applied coal tar ointments and progressively increasing doses of ultraviolet radiation (UVA and UVB) for treatment of psoriasis (Goeckerman regimen) and administration of a derivative of hematoporphyrin, in conjunction with exposure to red light, for treatment of certain kinds of tumors.[96, 97]

As more information becomes available about the interactions of light with drugs and with mammalian cells and tissues, the increasing potential for using these interactions to therapeutic advantage becomes evident. In the area of photochemotherapy, one approach under development is particularly interesting. It involves administration of a drug that is effective for treatment of a particular disease state, but that has been inactivated by adding a photochemically labile blocking group. Without light exposure, the drug, in theory, would exhibit little toxicity, but upon exposure to the appropriate wavelength of light in the specific target site, the drug would be "activated" and express its therapeutic potential.[98] Similarly, by applying concepts from photochemistry to synthetic chemistry, it may be possible to produce a whole spectrum of drugs or drug carriers (e.g., light-sensitive liposome systems[99]) that can be either activated or deactivated by exposure to specific wavelengths of electromagnetic radiation. Knowledge of the relative penetrance of various wavelengths of light to given tissues and/or application of fiber optic technology to expose tissues selectively to given wavelengths could then permit a significant degree of selective targeting to the desired site.

Ultraviolet radiation significantly affects the population distribution of circulating lymphocytes in humans[100, 101] and in experimental animals it has been shown to affect markedly a variety of immune responses including susceptibility to UV-induced tumor growth.[102] It has also been shown that in experimental animals, UVB exposure can result in suppression of photoallergy.[103] If rat islets of Langerhans are irradiated with UVB *in vitro*, prior to transplantation across a major histocompatibility barrier into diabetic rats, the survival of the grafts is enhanced.[104] Similarly, pre-exposure of an animal recipient of an allogenic bone marrow transplant to UVB has been shown to prevent development of graft-host disease.[105] These observations suggest enticing potential future uses of ultraviolet radiation in clinical practice. However, application of such techniques must await elucidation of the mechanisms by which UV affects immune responses and the scope of possible pathologic effects such responses may have.

References

1. U. Gallo and L. Santa Maria, Eds., *Research Progress in Organic, Biological and Medicinal Chemistry*, Vol. 3, North Holland Press, Amsterdam, 1972.
2. C. S. Foote, Photosensitized oxidations and singlet oxygen: Consequences in biological systems, In *Free Radicals in Biology*, Vol. 2, W. A. Pryor, Ed., Academic Press, New York, 1976.
3. T. B. Fitzpatrick, G. Szabó, N. Seiji, and W. C. Quevedo, Jr., Biology of the melanin pigmentary system, In *Dermatology in General Medicine*, T. B. Fitzpatrick, Ed., McGraw-Hill, New York, 1979.
4. L. C. Harber and D. R. Bickers, Photoprotection, In *Photosensitivity Diseases: Principles of Diagnosis and Treatment*, W. B. Saunders, Philadelphia, 1981.
5. M. A. Pathak, F. J. Riley, T. B. Fitzpatrick, and W. L. Cureven, Melanin formation in human skin induced by longwave ultraviolet and visible light, *Nature 193*:148, 1962.
6. J. H. Epstein, Photomedicine, In *The Science of Photobiology*, K. C. Smith, Ed., Plenum Press, New York, 1977.
7. J. A. Parrish, R. R. Anderson, F. Urbach, and D. Pitts, Immediate and short-term biologic effects of ultraviolet radiation on normal skin, In *UV-A: Biological Effects of Ultraviolet Radiation with Emphasis on Human Responses to Longwave Ultraviolet*, Plenum Press, New York, 1978.
8. M. A. Pathak and K. Stratton, A study of the free radicals in human skin before and after exposure to light, *Arch. Biochem. Biophys. 123*:468, 1968.
9. R. W. Gange and J. A. Parrish, Acute effects of ultraviolet radiation upon the skin, In *Photoimmunology*, J. A. Parrish, M. L. Kripke, and W. L. Morison, Eds., Plenum Press, New York, 1983.
10. W. C. Quevedo, Jr., T. B. Fitzpatrick, M. A. Pathak, and K. Simbow, Light and skin color, In *Sunlight and Man: Normal and Abnormal Photobiologic Responses*, M. A. Pathak, L. C. Harber, M. Seiji, and A. Kukita, Eds., University of Tokyo Press, Tokyo, 1974.
11. M. A. Pathak, K. Jimbow, and J. A. Parrish, Effect of UVA, UVB and psoralen on in vivo pigmentation. Cellular and subcellular characterization of delayed tanning induced by single or multiple exposures to UV-A, UV-B, or UV-8 and 8-methoxypsoralen, In *Pigment Cell*, V. Reley, Ed., Karger, Basel, 1976.

12. K. M. Kaidbey and A. M. Kligman, Sunburn protection by ultraviolet radiation-induced pigmentation, *Arch. Derm. 109*:200, 1978.
13. D. Cripps, Natural and artificial photoprotection, *J. Invest. Derm. 76*:154, 1981.
14. J. H. Epstein, K. Fukuyama, and K. Fye, Effects of ultraviolet radiation on the mitotic cycle and DNA, RNA and protein synthesis in mammalian epidermis in vivo, *Photochem. Photobiol. 12*:57, 1970.
15. W. L. Epstein, K. Fukuyama, and J. H. Epstein, Early effects of utraviolet light on DNA synthesis in human skin in vivo, *Arch. Derm. 100*:84, 1969.
16. H. P. Baden and C. Pearlman, The effect of ultraviolet light on protein and nucleic acid synthesis in the epidermis, *J. Invest. Derm. 43*:71, 1964.
17. J. H. Epstein, K. Fukuyama, and W. L. Epstein, UVL induced stimulation of DNA synthesis in hairless mouse epidermis, *J. Invest. Derm. 51*:445, 1968.
18. I. Willis, A. Kligman, and J. Epstein, Effects of long ultraviolet rays on human skin. Photoprotection or photoaugmentation? *J. Invest. Derm. 59*:416, 1972.
19. N. J. Lowe, A. K. Verma, and R. K. Boutwell, Ultraviolet light induced epidermal ornithine decarboxylase, *J. Invest. Derm. 71*:417, 1978.
20. A. K. Verma, N. J. Lowe, and R. K. Boutwell, Induction of mouse epidermal ornithine decarboxylase activity and DNA synthesis by ultraviolet light, *Cancer Res. 39*:1035, 1979.
21. D. M. Russell, Polyamines in growth, normal and neoplastic, In *Polyamines in Normal and Neoplastic Growth*, D. M. Russell, Ed., Raven Press, New York, 1973.
22. R. W. Gange and P. R. De Quoy, Topical spermine- and putrescine-stimulated epidermal DNA synthesis in the hairless mouse epidermis, *Brit. J. Derm. 103*:27, 1980.
23. W. M. Sams, Jr. and R. K. Winkelmann, The effect of ultraviolet light on isolated cutaneous blood vessels, *J. Invest. Derm. 53*:79, 1969.
24. J. C. Van der Leun, On the action spectrum of ultraviolet erythema, In *Organic, Biological and Medicinal Chemistry*, Vol. 3, Part 2, North Holland Press, Amsterdam, 1972.
25. L. C. Harber and D. R. Bickers, Sunburn (ultraviolet erythema), In *Photosensitivty Diseases: Principles of Diagnosis and Treatment*, W. B. Saunders, Philadelphia, 1981.
26. G. Logan and D. L. Wilhelm, Ultraviolet injury as an experimental method of the inflammatory reaction, *Nature 198*:968, 1963.
27. M. F. Sim, The response of mouse skin to ultraviolet irradiation and its modification by drugs, In *Proceedings of the Internatinal Symposium on Non-Steroidal Anti-Inflammatory Drugs*, S. Garattini and M. Dukes, Eds., Excerpta Medica, Amsterdam, 1964.
28. G. Logan and D. L. Wilhelm, Vascular permeability changes in inflammation. I. The role of endogenous permeability factors in ultraviolet injury, *Brit. J. Exp. Pathol. 47*:300, 1966.
29. T. S. Veninga and J. E. de Boer, Urinary excretion pattern of serotonin and 5-hydroxindole acetic acid in ultraviolet induced erythema, *J. Invest. Derm. 50*:1, 1968.
30. D. S. Snyder and W. H. Eaglstein, Intradermal antiprostaglandin agents and sunburn, *J. Invest. Derm. 62*:47, 1974.
31. W. H. Eaglstein and A. R. Marsico, Dichotomy in response to indomethacin in UV-C and UV-B induced ultraviolet light inflammation, *J. Invest. Derm. 65*:238, 1975.
32. W. L. Morison, B. S. Paul, and J. A. Parrish. The effects of indomethacin on long wave ultraviolet-induced delayed erythema, *J. Invest. Derm. 68*:130, 1971.
33. A. C. Alison, I. A. Magnus, and M. R. Young, Role of lysosomes and cell membranes in photosensitization, *Nature 209*:874, 1966.
34. B. E. Johnson and F. Daniels, Jr., Lysosomes and the reactions of skin to ultraviolet radiation, *J. Invest. Derm. 53*:85, 1969.
35. B. E. Johnson, F. Daniels, Jr., and I. A. Magnus, Response of human skin to ultraviolet light, *Photophysiology 4*:139, 1968.
36. W. M. Sams, Jr., J. G. Smith, and P. G. Burk, The experimental production of elastosis with ultraviolet light, *J. Invest. Derm. 43*:467, 1964.

37. K. Nakamura and W. C. Johnson, Ultraviolet light induced connective tissue changes in rat skin: A histopathologic and histochemical study, *J. Invest. Derm. 51*:253, 1968.
38. D. W. Owens and J. M. Knox, Influence of heat, wind and humidity on ultraviolet radiation injury, *Natl. Cancer Inst. Monogr. 50*:161, 1978.
39. J. A. Bain, H P. Rusch, and B. E. Kline, The effect of temperature upon ultraviolet carcinogenesis with wavelengths 2,800–3,400 Å, *Cancer Res. 3*:610, 1943.
40. F. Urbach, P. D. Forbes, and R. E. Davis, Modification of photocarcinogenesis by chemical agents, *J. Natl. Cancer Inst. 69*:229, 1982.
41. J. H. Epstein, Photocarcinogenesis, skin cancer and aging, *J. Amer. Acad. Derm. 9*:487, 1983.
42. P. D. Forbes, H. F. Blum, and R. E. Davies, Photocarcinogenesis in hairless mice: Dose response and influence of dose-delivery, *Photochem. Photobiol. 34*:361, 1981.
43. F. R. De Gruiji, J. B. van der Meer, and J. C. van der Leun, Dose-time dependency of tumor formation by chronic UV exposure, *Photochem. Photobiol. 37*:53, 1983.
44. H. S. Black, W. Lenger, M. MacCallum, and J. Gerguis, Relation of dietary lipid level to photocarcinogeneis, *Photochem. Photobiol. 37*:S39, 1983.
45. J. Scotto, T. R. Fears, and J. F. Fraumeni, Jr., Incidence of Non-Melanoma Skin Cancer in the United States, Publication no. (NIH) 82-2433, U.S. Department of Health and Human Services, 1981.
46. J. C. van der Leun, UV-carcinogenesis, *Photochem. Photobiol. 39*:861, 1984.
47. P. D. Forbes and R. E. Davies, Factors that influence photocarcinogenesis, in *Photoimmunology*, J. A. Parrish, M. L. Kripke, and W. L. Morison, Eds., Plenum Press, New York, 1983.
48. L. C. Harber, R. B. Armstrong, and M. Ichikawa, Current status of predictive animal models for drug photoallergy and their correlation with drug photoallergy in humans, *J. Natl. Cancer Inst. 69*:237, 1982.
49. P. S. Herman and W. M. Sams, Jr., Requirement for carrier protein in salicylamide sensitivity: The migration-inhibition test in contact photoallergy, *J. Lab. Clin. Med. 77*:572, 1971.
50. L. C. Harber, I. E. Kochevar, and A. R. Shalita, Mechanisms of photosensitization to drugs in man, In *Photomedicine*, J. Parrish, M. L. Kripke, and W. L. Morison, Eds., Plenum Press, New York, 1980.
51. O. F. Jillson and R. D. Baughman, Contact photodermatitis from bithionol, *Arch. Dermatol. 88*:409, 1963.
52. C. A. Ramsay, Skin responses to ultraviolet radiation in contact photodermatitis due to Fentichlor, *J. Invest. Derm. 72*:99, 1979.
53. K. H. Kaidbey and A. M. Kligman, Contact photoallergy to 6-methylcoumarin in proprietary sunscreens, *Arch. Derm. 114*:1709, 1978.
54. J. N. Burry, Photoallergies to Fentichlor and Multifungin, *Arch. Derm. 95*:287, 1967.
55. I. E. Kochevar, G. L. Zalar, J. Einbinder and L. C. Harber, Assay of contact photosensitivity to musk ambrette in guinea pigs, *J. Invest. Derm. 73*:144, 1979.
56. A. G. Motten, C. F. Chignell, and R. P. Mason, Spectroscopic studies of cutaneous photosensitizing agents—VI. Identification of the free radicals generated during the photolysis of musk ambrette, musk xylene and musk ketone, *Photochem. Photobiol. 38*:671, 1983.
57. P.-S. Song and K. J. Tapley, Jr., Photochemistry and photobiology of psoralens, *Photochem. Photobiol. 29*:1177, 1979.
58. R. S. Cole, D. Levithan, and R. R. Sinden, Removal of psoralen interstrand cross-links from DNA of *Escherichia coli*: Mechanism and genetic control, *J. Mol. Biol. 103*:39, 1976.
59. B. D. Goldstein and L. C. Harber, Erythropoietic protoporphyria: Lipid peroxidation and red cell membrane damage associated with photohemolysis, *J. Clin. Invest. 51*:892, 1972.
60. S. Lerman, Biological and chemical effects of ultraviolet radiation, In *Radiant Energy and the Eye*, Macmillan, New York, 1980.
61. S. Lerman, Biophysical aspects of corneal and lenticular transparency, In *Proceedings of the First International Symposium on Light and Oxygen Effects on the Eye*, S. D. Varma and S. Lerman, Eds., IRL Press, Ltd., Oxford, 1984.

62. S. Lerman, Observations on the use of high power lasers in ophthalmology, *IEEE J. Quantum Electronics, 20*:1465, 1984.
63. F. Fraunfelder, Ed., *Drug-Induced Ocular Side Effects and Drug Interactions*, Lea and Febiger, Philadelphia, 1982.
64. A. Potts and L. M. Gonasum, Toxicology of the eye, In *Toxicology*, J. J. Casarett and J. Doull, Eds., Macmillan, New York, 1975.
65. T. M. Cloud, R. Hakim, and A. C. Griffin, Photosensitization of the eye with methoxalen. I. Acute effect, *Arch. Ophthalmol. 64*:346, 1960.
66. T. M. Cloud, R. Hakim, and A. C. Griffin, Photosensitization of the eye. II. Chronic effects, *Arch. Ophthalmol. 66*:689, 1961.
67. R. G. Freeman and R. Troll, Photosensitization of the eye by 8-methoxypsoralen, *J. Invest. Derm. 53*:449, 1969.
68. S. Lerman and R. Borkman, A method for detecting 8-methoxypsoralen in the ocular lens, *Science 197*:1287, 1977.
69. S. Lerman, M. Jocoy, and R. Borkman, Photosensitization of the lens by 8-methoxypsoralen, *Invest. Ophthalmol. Vis. Sci. 16*:1065, 1977.
70. J. G. Jose and K. L. Yielding, Unscheduled DNA synthesis in lens epithelium following ultraviolet irradiation, *Exp. Eye Res. 24*:113, 1977.
71. J. G. Jose and K. L. Yielding, Photosensitive cataractogens, chlorpromazine and methoxypsoralen cause DNA repair synthesis in lens epithelial cells, *Invest. Ophthalmol. Vis. Sci. 17*:687, 1978.
72. S. Lerman, J. Megaw, and I. Willis, Potential ocular complications of PUVA therapy and their prevention, *J. Invest. Derm. 74*:197, 1980.
73. S. Lerman, J. Megaw, and I. Willis, The photoreaction of 8-MOP with tryptophan and lens proteins, *Photochem. Photobiol. 31*:235, 1980.
74. J. Megaw, J. Lee, and S. Lerman, NMR analyses of tryptophan-8-methoxypsoralen photoreaction products, *Photochem. Photobiol. 32*:265, 1980.
75. M. N. Crylin, A. Pedirs-Leftick, and J. Sugar, Cataract formation in association with ultraviolet photosensitivity, *Ann. Ophthalmol. 12*:786, 1980.
76. S. Lerman, J. Megaw, K. Gardner, Y. Takei, and I. Willis, Localization of 8-methoxypsoralen in ocular tissues, *Ophthalmic Res. 13*:106, 1981.
77. H. C. Wulf and M. P. Andreasen, Distribution of ^{3}H-8-MOP and its metabolites in rat organs after a single oral administration, *J. Invest. Derm. 76*:252, 1981.
78. H. C. Wulf and M. P. Andreasen, Concentration of ^{3}H-8-methoxypsoralen and its metabolites in the rat lens and eye after a single oral administration, *Invest. Ophthalmol. Vis. Sci. 22*:32, 1982.
79. S. Lerman, Ocular phototoxicity and PUVA therapy: An experimental and clinical evaluation, *J. Natl. Cancer Inst. 69*:287, 1982.
80. S. Lerman, J. Megaw, and K. Gardner, P-UVA therapy and human cataractogenesis, *Invest. Ophthalmol. Vis. Sci. 23*:801, 1982.
81. S. Lerman, Y. Takei, Y. Franks, J. Megaw, K. Gardner, and A. Gammon, Photobinding of ^{3}H-8-methoxypsoralen to monkey intraocular tissues, *Invest. Ophthalmol. Vis. Sci. 23* (Suppl.):200, 1983.
82. J. Parrish, T. B. Fitzpatrick, L. Tanenbaum, and M. A. Pathak, Photochemotherapy of psoriasis with oral methoxalen and longwave ultraviolet light, *New Engl. J. Med. 291*:1207, 1974.
83. J. A. Parrish, T. B. Fitzpatrick, C. Shea, and M. A. Pathak, Photochemotherapy of vitiligo. Use of orally administered psoralens and a high intensity longwave ultraviolet light (UV-A) system, *Arch. Derm. 112*:1531, 1976.
84. P. Dayhaw-Barker and F. M. Barber, II, Retinal effects of short term exposure to 8-MOP and UV-A, *Photochem. Photobiol. 37*(Suppl.):S83, 1983.
85. P. Dayhaw-Barker, F. M. Barber, II, and K. Diebert, Effects of three drugs on the retinal threshold to damage, *Amer. J. Opt. Phys. Optics 59*:10, 1982.
86. S. I. Davidson, Reports of ocular adverse reactions, *Trans. Ophthalmol. Soc. U.K. 93*:495, 1973.

87. S. Lerman, J. Megaw, and K. Gardner, Allopurinol and human cataractogenesis, *Amer. J. Ophthalmol. 94*:141, 1982.
88. S. Lerman, F. Fraunfelder, and J. Megaw, Further studies of allopurinol in human cataractogenesis, *Amer. J. Opthalmol. 97*:205, 1984.
89. D. S. Hull, S. Csukas, and K. Green, Trifluoperazine: Corneal endothelial phototoxicity, *Photochem. Photobiol. 38*:425, 1983.
90. F. Kretzner and R. Mehta, An age-dependent adriamycin-induced photoreceptor atrophy in albino rats, *Invest. Ophthalmol. Vis. Sci. 20* (Suppl.):40, 1981.
91. A. T. Johnson and F. Kretzner, Photoreceptor degeneration induced by adriamycin, *Invest. Opththalmol. Vis. Sci. 24* (Suppl.):181, 1983.
92. J. H. Epstein, Chemical Phototoxicity in humans, *J. Natl. Cancer Inst. 69*:265, 1982.
93. J. Cadet, L. L. Voituriez, F. Gaboriau, P. Vigny, and S. Della Negra, Characterization of photocycloaddition products from reaction between thymidine and the monofunctional 3-carbethoxypsoralen, *Photochem. Photobiol. 37*:363, 1983.
94. J. E. Hearst, Psoralen photochemistry, *Annu. Rev. Biophys. Bioeng. 10*:69, 1981.
95. M. S. Ashwood-Smith and E. L. Grant, Chromosome damage produced by psoralen and ultraviolet light, *Brit. Med. J. 1*:342, 1976.
96. T. Dougherty, R. E. Thoma, D. G. Boyle, and K. R. Weishaupt, Interstitial photoradiation therapy for primary solid tumors in pet cats and dogs, *Cancer Res. 41*:401, 1981.
97. I. A. Menon, S. Persad, H. F. Haberman, C. J. Kurian, and P. Basu, A qualitative study of the melanins from blue and brown human eyes, *Exp. Eye Res. 34*:531, 1982.
98. K. C. Smith, Photobiology and photomedicine: The future is bright, *J. Invest. Derm. 77*:2, 1981.
99. J. A. Parrish, New concepts in therapeutic photomedicine: Photochemistry, optical targeting and the therapeutic window, *J. Invest. Derm. 77*:45, 1981.
100. P. Hersey, G. Haran, E. Hasic, and A. Edwards, Alteration of T cell activity in normal subjects after exposure to sunlight, *J. Immunol. 31*:1, 1983.
101. P. Hersey, E. Hasic, A. Edwards, M. Bradley, G. Haran, and W. H. McCarthy, Immunological effects of solarium exposure, *Lancet 1*:545, 1983.
102. W. L. Morison, Photoimmunology, *Photochem. Photobiol. 40*:781, 1984.
103. R. D. Granstein, W. L. Morison, and M. L. Kripke, The role of suppressor cells in the induction of murine photoallergic contact dermatitis and in its suppression by prior UVB irradiation, *J. Immunol. 130*:2099, 1983.
104. H. Lau, K. Reetsma, and M. A. Hardy, Prolongation of rat islet allograft survival by direct ultraviolet irradiation of the graft, *Science 223*:607, 1984.
105. A. Glazier, W. L. Morison, C. Bucana, A. D. Hess, and P. J. Tutschka, Suppression of epidermal graft-versus-host disease with ultaviolet radiation, *Transplantation 37*:211, 1984.

2

PHOTOTOXICITY

KARL L. GABRIEL, V.M.D., Ph.D.
Biosearch, Incorporated
Philadelphia, Pennsylvania

P. DONALD FORBES, Ph.D.
RONALD E. DAVIES, Ph.D.
Center for Photobiology
Skin and Cancer Hospital
Temple University Health Sciences Center
Philadelphia, Pennsylvania

Abstract

An overview of the field of phototoxicity is presented, drawing on the more recent literature and discussing light sources, *in vitro* models, animal models, and human experimental methods currently being employed.

Phototoxicity is an abnormal cutaneous response to ultraviolet or visible radiation associated with the presence of a chemical, applied either locally or systemically. Like primary irritancy and unlike photoallergy, the phototoxic response is elicited on the first exposure if the conditions of exposure are sufficiently severe. This paper considers only phototoxicity; photoallergy is described in the following article. The word "photosensitization" has been used to describe both these photobiological reactions; the word "photosensitizer" also has a specific connotation in photochemistry.[1] We will avoid the nonspecific use of "photosensitization" in this paper.

There are at least three different types of phototoxic responses; a particular chemical may elicit one or more of these responses. The first type involves little or no immediate response but is characterized by a strong and often prolonged delayed erythema and edema; onset is generally within 8–24 hr and the response may last for several days. It may resemble ordinary sunburn but the erythema may be a darker red. For a given phototoxic agent the response is dependent on total radiation dose, with little dependence on intensity. The prototype chemical for this type of reaction is psoralen.

The second type of response involves a much more rapid, usually transient, erythema that may be apparent immediately after exposure and is usually gone within 1 or 2 days. Edema is not usually observed. This response is characterized by a localized burning sensation that is dependent on radiation intensity, can be severe or intolerable, and disappears when exposure is terminated. It is not clear whether the erythema, like the pain, depends on radiation intensity, although it does appear to depend on radiation dose. The prototype chemical for this type of reaction is anthracene.

The third type of response involves the production of wheals, with or without accompanying flares. The response is typically rapid and can be quite transient. It is associated with a burning sensation that is dependent on the radiation intensity and can involve severe tissue destruction. The prototype chemical for this type of reaction is hematoporphyrin.

The choice of radiation sources for various phototoxicity studies is dictated by the following considerations:

1. The emission spectrum should include the radiation most likely to elicit the reaction under study. Where reliable predictions of appropriate spectral regions are not available, spectral coverage should be as broad as possible, either from broad-spectrum sources or from several sources, in independent tests.
2. The radiation dose should be as high as is practically achievable within the following restrictions.
   a. Delivered dose should produce little or no apparent damage in normal, untreated skin.
   b. Intensity should be limited to avoid appreciable local temperature rises, leading to pain or injury.
   c. Delivery time must be consistent with the subject's comfort as well as practical limitations of time and number of subjects in clinical studies.
   d. Uniformly irradiated area must be sufficient for purposes of the test; area–intensity tradeoffs may be possible within the restrictions noted above.
   e. The irradiation source must be reliable, reasonably simple to operate, and not so exotic that alternatives or back-up would be difficult to obtain.
   f. Nonradiative characteristics of the source (electrical or thermal hazards, source–subject geometry) must be suitable for safe, convenient, and reproducible operation.

Light Sources

Many phototoxic agents are activated by light in the 320–400-nm spectral region. A convenient source of this radiation is the "blacklight" fluorescent lamp. Forbes[2] used a fixture containing four F40BL lamps, which provided a broad band output in the UVA region centered at 350 nm. A 6-mm thick sheet of glass interposed between the source and the skin eliminated the small amount of UVB emitted by the lamps. The source was used to identify phototoxic samples among fragrance raw materials, as described below.

Since the biological action spectrum for a particular, new test agent is unknown, exposure to a broad spectrum of wavelengths is essential for screening purposes. To reduce the likelihood of missing agents activated by visible light (anthroquinone dyes, for example), studies should be conducted with visible light as well as UV sources. Consideration should also be given to the possibility of a phototoxic reaction stimulated by radiation < 320 nm, although the response may be difficult to distinguish from directly induced erythema.

Kaidbey and Kligman[3] discussed the ideal light source for phototoxicity studies; it should be easy to maintain and have a dependable, constant output with high intensity in the UV and visible regions, and highly erythemogenic wavelengths below 320 nm would be filtered out. The xenon-arc solar simulator is a good source for this type of work. In a device described by Berger,[4] the lamp is equipped with a UV-reflecting dichroic mirror that reduces the amount of heat. The visible light output can be enhanced by changing the dichroic mirror.

Cole[5] points out the need to exercise caution when purchasing blacklight fluorescent lamps for use in photobiology studies and noted a difference of spectral distribution in blacklight lamps. The coding of the lamps was identical and it was impossible to differentiate two spectrally different sources based on labeling or packaging materials. This paper points out the need to check source outputs and intensities carefully before using them in research on living systems.

Plewig[6] discussed a light source consisting of a 200 W super-pressure mercury lamp and a flexible light guide (a tube with a reflective inner coating and a light conducting liquid core). With proper filtration, the UVB was cut off. The output of the filtered guide was centered mainly at 366 nm.

Cole[7] used a liquid light guide leading from a xenon arc solar simulator. The spectral composition of the solar simulator UVR output was unchanged by the light guide. Other types of apparatus, utilizing the liquid light guide for final delivery of the radiation, will probably be available in the future since the liquid light guide is a convenient way of delivering radiation to the biological test system.

Human Phototoxicity Studies

The ethical considerations involved in conducting experimental studies in human subjects are as pertinent in the area of phototoxicity as in any other area of research;

nevertheless, the limited risks associated with skin testing appear to justify certain procedures (see examples below). The induction of phototoxicity is not known to produce a persistent, long-lasting alteration in the subject. Aside from some residual pigmentation, threshold phototoxic reactions in a small area are mild and transient. Therefore the use of a positive control material such as 8-methoxypsoralen (8-MOP) may be considered to assure the investigator that the methodology being employed is adequate to detect at least this type of phototoxic agent.

Kaidbey and Kligman[3] discuss phototoxicity testing in human subjects. They enumerate several materials that they have demonstrated to be phototoxic in human subjects. Among these are 7-methoxycoumarin, dimethylanthranilate, coal tars and coal tar distillates, and the psoralens (8-methoxy-, 5-methoxy-, and trimethylpsoralen). These same authors, along with Webster[8] demonstrated phototoxicity in humans of the nonsteroidal anti-inflammatory drug, benoxaprofen. Kligman and Kaidbey[9] describe the human subject as "an appropriate and sensitive animal" for identifying photosensitizing drugs administered either topically or systemically.

Dielte[10] studied coal tar phototoxicity in 32 subjects. They found that the two major manifestions of coal tar phototoxicity—reduced delayed erythema threshold and susceptibility to the smarting reaction—persisted for at least 30 hr after tar removal.

Kaidbey and Kligman[11] showed that with certain poorly penetrating substances, such as water-soluble dyes, applications to scarified skin were necessary to produce phototoxicity. They also studied angelica root oil, cumin oil, expressed bergamot oil, oil orange bitter (West India), lime oil (expressed), rue oil, bergamot DXD (twice rectified), and lemon oil. All were phototoxic when applied to normal, human skin except twice-rectified bergamot oil. Several halogenated phenolic compounds were also studied. Tetrachlorosalicylanilide (TCSA), bithionol, and jadit, an antifungal agent, were phototoxic.

Animal Models

Phototoxicity has been studied in several species with normal or sparse hair coats: mice, rabbits, guinea pigs, and swine. Certain aspects have even been investigated in the squirrel monkey and the hamster. There are variations in response among these species. The hairless mouse and the guinea pig are the most widely used test models. The following is a description of a method used by the authors. The experimental material may be administered either topically or systemically. Radiation is delivered and the animals are retained for observation, restrained and unanesthetized. Treatment conditions are designed to facilitate distinctions among effects of occlusion, thermal load, primary irritancy, and phototoxicity. Two radiation sources are used, one to maximize radiation load in the most common spectral region for phototoxicity, the other to provide a broad range of radiation for detection of unusual responses. Both are to be used at intensities chosen to produce maximum "no-effect" doses in reasonable exposure periods. Direct or indirect monitoring of radiation load, to provide uniform and reproducible test conditions, is conducted. Both positive and negative controls are employed to provide direct demonstration of test adequacy and to detect any uncontrolled influences on test sensitivity.

Test conditions and response parameters should be selected to provide unambiguous threshold assessments that do not entail subjective estimates of response magnitude. It is also a good idea to incorporate a defined dilution of the test material into the testing procedure to provide an abbreviated severity assay and to guard against photobiological anomalies such as self-shielding. Finally a simple, straightforward, enumerative statistical technique should be used to provide a high level of confidence in cases of zero responses.

Forbes[2] demonstrated that responses of mouse skin to known phototoxic agents are similar to responses in humans or swine, except that mouse skin requires less of the phototoxic agent to elicit symptoms. These workers studied 160 fragrance raw materials on hairless mouse skin and on swine skin followed by exposure to a solar simulator or near-ultraviolet light. Twenty-one materials elicited a phototoxic response. When several selected materials were tested in sunlight and under three laboratory light sources on the skin of mice, humans, and miniature swine, mouse skin was the most sensitive; reactions in human and swine skin were qualitatively and quantitatively similar.

In addition to using dorsal back skin as the site for eliciting the phototoxic response, Stott,[12] Sambuco and Forbes,[13] and Cole[7] have evaluated the mouse ear for phototoxic effects. Measurements of ear thickness are taken using a hand-held micrometer before and after treatment with the experimental chemical and the light source. This method is objective, nondestructive, and quantifiable. The phototoxic response in the ears was measurable over an extended period of time and repeated measurements did not alter the response.

Guillot[14] describes a method using guinea pigs to assess both the phototoxic and photoallergic potentials of a substance. The substance is applied under occlusive patches using Freund's complete adjuvant during the photoallergy induction to create immunostimulation.

In Vitro Methods

Excellent information on sites and mechanisms of phototoxic reactions is presented by Kornhauser.[15]

In 1965, Daniels[16] published a simple microbiological method for demonstrating phototoxicity (inhibited growth of *Candida albicans* on agar plates). He used his method to demonstrate the phototoxicity of several classes of compounds. Weinberg and Springer[17] used bakers' yeast (*Saccharomyces*) in a similar technique. They reported detection levels of 8-MOP as the standard positive agent and they confirmed the *in vitro* phototoxicity of several fragrance raw materials.

Freeman[18] exposed Hep-2 cells in tissue culture to nontoxic concentrations of chemicals, subsequently exposed the cells to light sources, and was able to demonstrate a cytotoxic effect indicating phototoxicity with 8-MOP, hematoporphyrin, pseudourea, cholorquine, thioridazine, and coal tar. Sulfanilamide, which is associated with immunologic photoreactions, did not give positive results in this system. Another *in vitro* technique involved photohemolysis of erythrocytes. Hemolysis occurred promptly with pseudourea, coal tar, chlorpromazine, demethylchlortetracycline, and chloroquine; neither sulfanilamide nor 8-MOP induced photohemolysis.

Webster and co-workers[8] studied the phototoxic activity of benoxaprofen (2-[4-chlorophenyl]-α-methyl-5-benzoxazole acetic acid), a nonsteroidal anti-inflammatory drug. *In vitro* studies demonstrated that benoxaprofen and UV irradiation produce a dose-dependent lysis of sheep erythrocytes, which did not appear to be dependent on the presence of oxygen. Human neutrophils were lysed by the drug and UV light, which resulted in the liberation of both cytoplasmic and liposomal enzymes.

The authors suggested that phototoxic urticaria provoked by benoxaprofen may be due to a direct photolytic effect on human dermal mast cells.

Ashwood-Smith[19] describes a photobiological assay based upon inhibition of growth in the DNA-repair-deficient bacterium *E. coli* B_{S-1} for studying possible photosensitizers. Use of an organism deficient in normal DNA repair increased the sensitivity of detection over methods in previously published papers, according to the authors.

Morison[20] exposed human lymphoid cells to UVA (320–400 nm) and in some cases UVB (280–320 nm) radiation in the presence of various chemicals and measured ^{3}H-thymidine uptake into nuclear DNA. Eleven chemicals, reported to be photosensitizers in humans, were tested with this method. Nine were phototoxic in combination with UVA radiation and the two that were not, were found to be phototoxic with UVB radiation. Retinoic acid, which is not reported to be a photosensitizer, but does potentiate experimental photocarcinogenesis and inhibit UVB-induced alterations of DNA synthesis in mouse skin, was found to be phototoxic in combination with UVA radiation.

In vitro methods for studying phototoxicity are important because they are usually rapid and inexpensive and therefore allow one to screen more chemicals. The methods employing most microorganisms are not very specific and produce many false-positive reactions, compared with *in vivo* tests with animals or humans. More complex systems, utilizing DNA-repair-deficient mutants, may be useful in particular situations and may provide information on the site of mechanism or action. The use of *in vitro* methods should add a great deal to the evolution of this field when the relationships to data from *in vivo* studies are better understood.

References

1. J. D. Spikes, Photosensitized reactions in mammals, In *Experimental and Clinical Photoimmunology*, R. A. Daynes and J. D. Spikes, Eds., CRC Press, Boca Raton, Florida, 1983.

2. P. D. Forbes, F. Urbach, and R. E. Davies, Phototoxicity testing of fragrance raw materials, *Fd. Cosmetic Toxicol.* *15*(1):55, 1977.
3. K. H. Kaidbey and A. M. Kligman, Identification of contact photosensitizers by human assay, In *Current Concepts in Cutaneous Toxicity*, V. A. Drill and P. Lazar, Eds., Academic Press, New York, 1980.
4. D. S. Berger, Specification and design of solar ultraviolet simulators, *J. Invest. Derm.* *53*:192, 1969.
5. C. A. Cole, P. D. Forbes, and R. E. Davies, Different biological effectiveness of blacklight fluorescent lamps available for PUVA therapy, *J. Amer. Acad. Derm.* *11*:599, 1984.
6. G. Plewig, C. Hofmann, O. Braun-Falco, G. Nath, and A. Kreitmair, A new apparatus for the delivery of high intensity UVA and UVA + UVB irradiation, and some dermatological applications, *Brit. J. Derm.* *98*:15, 1978.
7. C. Cole, C. Sambuco, P. Forbes, and R. Davies, Response to ultraviolet radiation: Ear swelling in hairless mice, *Photodermatology* *1*:114, 1984.
8. G. F. Webster, K. H. Kaidbey, and A. M. Kligman, Phototoxicity from benoxaprofen: *In vivo* and *in vitro* studies, *Photochem. Photobiol.* *36*:59, 1982.
9. A. M. Kligman and K. H. Kaidbey, Human models for identification of photosensitizing chemicals, *J. Natl. Cancer Inst.* *69*(1):269, 1982.
10. K. M. Dielte, R. W. Gange, R. S. Stern, K. A. Arndt, and J. A. Parrish, Coal tar phototoxicity: Kinetics and exposure parameters, *J. Invest. Derm.* *81*:347, 1983.
11. K. H. Kaidbey and A. M. Kligman, Identification of topical photosensitizing agents in humans, *J. Invest. Derm.* *70*:149, 1978.
12. C. W. Stott, J. Strasse, R. Bonomo, and A. H. Campbell, Evaluation of the phototoxic potential of topically applied agents using long-wave ultraviolet light, *J. Invest. Derm.* *55*(6):335, 1970.
13. C. P. Sambuco and P. D. Forbes, Quantitative assessment of phototoxicity in the skin of hairless mice, *Fd. Chem. Toxic.* *22*(3):233, 1984.
14. J. P. Guillot, J. F. Gonnet, and J. F. Loquerie, New method for the assessment of phototoxic and photoallergic potentials by topical applications in the albino guinea pig, *J. Toxicol. Cut. Ocular Toxicol.* *4*(2):117, 1985.
15. A. Kornhauser, W. Wamer, and A. Giles, Jr., Light-induced dermal toxicity: Effects on the cellular and molecular level, In *Dermatotoxicology*, 2nd Ed., F. N. Marzulli and H. L. Maibach, Eds., Hemisphere, Washington, D.C., 1983.
16. F. Daniels, Jr., A simple microbiological method for demonstrating phototoxic compounds, *J. Invest. Derm.* *44*:259, 1965.
17. E. H. Weinberg and S. T. Springer, The evaluation *in vitro* of fragrance materials for phototoxic activity, *J. Soc. Cosmetic Chem.* *32*:303, 1981.
18. R. G. Freeman, Interaction of phototoxic compounds with cells in tissue culture, *Arch. Derm.* *102*:521, 1970.
19. M. J. Ashwood-Smith, G. A. Poulton, O. Ceska, M. Lin, and E. Furniss, An ultrasensitive bioassay for the detection of furocoumarins and other photosensitizing molecules, *Photochem. Photobiol.* *38*(1):113, 1983.
20. W. L. Morison, D. J. McAuliffe, J. A. Parrish, and K. J. Bloch, *In vitro* assay for phototoxic chemicals, *J. Invest. Derm.* *78*:460, 1982.

3

FUNDAMENTAL CONCEPTS IN PHOTOIMMUNOLOGY AND PHOTOALLERGY

THOMAS J. STEPHENS, Ph.D.
Mary Kay Cosmetics, Inc.
Dallas, Texas

PAUL R. BERGSTRESSER, M.D.
Department of Dermatology
University of Texas Health Science Center
Dallas, Texas

Abstract

Exposure to sunlight, particularly ultraviolet light (UV), can produce changes in the status of the immune system. Photoimmunology is the study of the distortions in "normal" immune responses that may be produced by UV and/or visible light exposure. Photoallergy, on the other hand, is an acquired immunologically mediated reactivity to a chemical or drug precipitated by the formation of photoproducts. This article reviews the key principles with supporting evidence from both photoimmunology and photoallergy. Reviews of the fundamental concepts of cutaneous structure and function, inflammation, immunology, and allergic contact dermatitis are presented as introduction. Subsequent sections discuss the capacity of UV to alter allergic contact dermatitis. The second part of the article presents an overview of past clinical observations and important laboratory animal studies relevant to photoallergy, including new techniques for predictive testing for photoallergic contact dermatitis in animals and helpful suggestions for handling problems when assessing consumer products for photoallergenicity. The article concludes by presenting approaches to diagnosing patients with suspected photoallergic contact dermatitis and photosensitivity to chemicals and drugs.

Introduction

Sunlight plays a vital role in our lives. Through photosynthesis, plants are able to capture solar energy and convert it into chemical energy which is then converted metabolically in humans and other animals into biomolecules and energy. However, our dependence on sunlight is not limited to the growing of food. Seasons and diurnal rhythms are also determined by cycles of sunlight. Physicians have used natural sunlight to treat both rickets and tuberculosis. In many societies, we associate good health with sun-induced rosy cheeks and tanned skin, even though scientists have established a link between sun exposure and both premature aging and skin cancer.[1] Sunlight can also affect responses of the immune system to antigens; the exploration of this relationship requires an understanding of photobiology.

Sunlight or solar energy consists of a spectrum of electromagnetic radiation. It arises in the sun from a nuclear fusion of hydrogen atoms to form helium atoms and electrons producing a core temperature[2] of 20,000,000 °K and a surface temperature of 5,400 °K. Newton, in the late 1660s, purchased glass prisms at a fair to investigate the phenomenon of light. He observed that sunlight passing through a prism will form a continuous strip of color images, with violet at the top and red at the bottom. He called this sequence of images a color spectrum and termed a single color in the spectrum as monochromatic. Newton indicated that light is not colored; the sensation of color is perceived by the human mind. In 1801, an English physician, Thomas Young, described light as a wave and thus explained certain color effects observed previously by Newton. Through the application of instrumentation rather than the human eye, 19th century scientists extended the range of the solar spectrum beyond visible light and quantitated the energy associated with both ultraviolet and infrared regions.[3]

Electromagnetic radiation consists of a composite of wavelengths, varying from the high-energy regions, including gamma, X, and vacuum ultraviolet radiation to the low-energy range of millimeter, micrometer, and radio radiation.[3] The energy of electromagnetic radiation is measured as photons or bundles of light energy, and photons in each wavelength of light are associated with different energy levels. For example, photons of 400-nm wavelength have twice the energy[4] of those 800 nm in wavelength. Mathematically, this relationship between energy and wavelength of light is given by the formula:

$$E = N\,h\nu = \frac{C}{\lambda} = \frac{286{,}000}{\lambda}\ \text{kcal}$$

where E is the energy in kilocalories of a mole or Einstein of photons; N is 6.023×10^{23}, Avogadro's number; ν is the frequency in vibrations per second; h is Planck's constant (1.58×10^{-34} cal sec); C is the velocity of light (186,000 miles/sec or 3×10^{10} cm/sec); and λ is the wavelength in angstroms.[2,4,5]

The total amount of solar energy reaching the earth's surface[2,6] is approximately 10^{24} cal per year. Wavelengths less than 290 nm are absorbed by the atmosphere including stratospheric ozone, and do not reach the earth's surface. In the biologically active area of the spectrum of solar energy (290–700 nm), large amounts reach the surface. Visible light comprises the wavelengths between 400 and 700 nm and appears as the spectrum of familiar colors, red through violet. This portion of the solar spectrum causes the human brain to recognize an event by passing through the eye and stimulating the retina. Visible light also penetrates skin to reach subcutaneous structures, but this penetration is uncommonly associated with injury.[7]

Wavelengths in the 200–400-nm range of the energy spectrum, called ultraviolet radiation (UV), have been divided conveniently into three regions, UVC (200–290), UVB (290–320 nm), and UVA (320–400 nm). UVC, or far UV, induces erythema in normal skin and causes photokeratitis (inflammation of the cornea and conjunctivitis).[7] UVC is also used in the food industry to minimize contamination by microorganisms, and thus has been called germicidal radiation. UVB, also called middle UV, represents only approximately 1.5% of the solar energy reaching the earth's surface,[8] but its important biological effects include mutagenesis in bacteria, skin cancer, and sunburning. Long-term exposure to UVB is also thought to be responsible for many of the changes in human skin associated with aging.[9, 10]

In contrast, UVA was thought for many years to be innocuous because of its relative inefficiency in producing skin erythema and its poor germicidal properties.[8] Since the development of high-intensity monochromatic ultraviolet light sources, UVA has also been shown to have important biological effects. Prolonged exposure to high-intensity UVA is both erythemogenic (producing redness) and melanogenic (tanning).[7, 9] Additionally, UVA has been proposed to cause certain types of cataracts as well as potentiating the biological effects of UVB exposure.[7]

In industry, UVA is used to photopolymerize and to induce the photochemical reactions that alter latex, rubber, glass, metal, paper, and photographs. UVA has also played an important role in the treatment of disease. Photochemotherapy is the therapeutic combination of drug and light to treat cutaneous diseases, such as PUVA: the combination of a photosensitive drug (*p*soralen) and *UVA* radiation. Psoralens are naturally occurring tricyclic furocoumarin compounds activated by UVA to produce phototoxicity. Activation of psoralens leads to the formation of thymine–psoralen photoproducts that inhibit DNA synthesis. Psoralens can be delivered to skin by either topical or systemic administration, and this treatment has been used successfully to treat such dermatologic conditions as vitiligo (acquired cutaneous amelanosis), psoriasis (increased proliferation of epidermis), mycosis fungoides (cutaneous T cell lymphoma), and atopic dermatitis (a relatively common form of eczema).[8, 10]

Another medical application of UVA is the Wood's light, which contains a medium-pressure mercury arc and a nickel oxide glass filter that emits UVA light. Dermatologists use the Wood's light to examine the skin for microorganisms that produce fluorescent by-products, such as *Pseudomonas* infection, erythrasma, and certain types of dermatophytes.[8]

Exposure to sunlight, particularly UV, can produce alterations in the status of the immune system. Photoimmunology is the study of distortions in "normal" immune responses produced by exposure to UV and/or visible light. Photoallergy, on the other hand, is an acquired immunologically mediated reaction requiring the formation of a chemically reactive photoproduct.[11] This review surveys the key principles and supporting experimental evidence defining the fields of photoimmunology and photoallergy.

As a starting point, we provide a series of key terms and definitions. Several of these terms are then reviewed in greater detail in subsequent sections. We will first review principles of cutaneous structure and function, inflammation, immunology, and allergic contact dermatitis, and use these as the basis for our discussion of the capacity of UV to alter allergic contact dermatitis.

In the second half, the readers are provided with an overview of clinical observations and important experimental studies relevant to photoallergy. Supplementing the discussion of photoallergy is a short presentation of recent developments in predictive testing for photoallergic dermatitis (photo-ACD) in animal models as well as some helpful suggestions for handling problems when assessing consumer products for photoallergenicity. The final section of this article presents clinical approaches employed by dermatologists in diagnosing patients with photoallergic contact dermatitis and photosensitivity to chemicals and drugs.

Terminology

Inflammation

A reproducible change in a tissue or organ characterized by redness (erythema), swelling (edema), warmth, pain, and an influx of leukocytes.

Dermatitis

Inflammation in skin. Acute dermatitis presents with redness, swelling, warmth, and pain, and it may be associated with the appearance of superficial blisters. A distinction is made between these objective irritation manifestations and subjective irritation manifestations such as burning, itching, or stinging reactions.

Immunology

A scientific discipline that studies the acquisition and regulation of protective (destructive) inflammatory reactions against "foreign" substances. An important issue in the regulation of immunity is the capacity to differentiate between "self" and "nonself."

Contact Dermatitis

Acute (or chronic) inflammatory reaction resulting from the application of substances to the skin surface. The distribution of contact dermatitis is limited to sites of exposure and it consequently may exhibit unusual, but characteristic, patterns across skin surfaces. Although two types predominate—irritant dermatitis and allergic contact dermatitis—they have many common features. On physical examination, contact irritant and allergic contact dermatitis are frequently indistinguishable; some materials have the capacity to produce both irritant and allergic reactions.

Irritant Contact Dermatitis

Direct injury from a chemical, biological, or physical insult. It requires no previous exposure to the irritant and will occur in virtually all individuals who are exposed. Two chemicals that produce typical contact irritant reactions are sodium lauryl sulfate and benzene.

Allergic Contact Dermatitis

An acquired, immunologically mediated, inflammatory reaction that will occur in a limited percentage of exposed individuals. Each chemical sensitizer is defined as a hapten, since it must bind chemically to proteins in skin before becoming immunologically reactive (allergen). Typical sensitizers (haptens) for contact allergic reactions include the pentadecylcatechol found in poison ivy and dinitrochlorobenzene (DNCB).

Contact Hypersensitivity

An animal model for allergic contact dermatitis as it occurs in humans.

Contact Sensitivity

Contact hypersensitivity.

Delayed-Type Hypersensitivity

A delayed and immunologically mediated inflammatory reaction attributed primarily to the cellular immune response.

Photoimmunology

The study of distortions in "normal" immune responses that may be produced by ultraviolet or visible electromagnetic radiation. Primary attention has been paid to electromagnetic radiation effects on the skin and eyes.

Photosensitivity

A broad term used to describe an adverse reaction to light, which may be phototoxic or photoallergic in nature.

Phototoxicity

The conversion of an otherwise nontoxic chemical to one directly toxic to skin after the absorption of electromagnetic radiation. Prototypes are tetracycline and psoralens.

Photoallergy

An acquired immunologic reactivity dependent on antibody or cell-mediated hypersensitivity. Photoallergy is observed in two forms: erythema with urticaria and acute dermatitis with eczema.

Hapten

A relatively small chemical which, by itself, is unable to induce an immune response, but which forms a complete antigen or allergen when bound chemically to tissue.

Cutaneous Structure and Function

The normal anatomical structure of skin observable by light microscopy is relatively constant on most body surfaces. An outer layer or compartment, called epidermis, consists of a thin cellular matrix, and in humans contains three varieties of cells: keratinocytes, melanocytes, and Langerhans cells (LC). Human epidermis averages 50 μm in thickness and it contains a surface density of approximately 50,000 viable cells/mm^2. With the light microscope, the three cell types that make up human epidermis become relatively indistinguishable, but substantial differences become apparent when specialized histochemical techniques are used, or when these cells are tested for their functional attributes. The most prominent function of epidermis is due to the laminated surface aggregate of differentiated keratinocytes, the stratum corneum, which serves as a physiological barrier to chemical penetration and microbiological invasion from without as well as fluid loss from within. Although keratinocytes make up 95% of epidermal cells, the other populations exert functions far more important than their relatively small numbers might suggest. Melanocytes elaborate the radiation-absorbing pigment, melanin, which plays a major role in protecting skin from ultraviolet radiation. Langerhans cells, representing only 3% of epidermal cells, play the critical immunologic role of antigen presentation during the acquisition of contact hypersensitivity and photoallergic contact dermatitis. Thus, the function of Langerhans cells will figure prominently in this review.

Lower in skin and beneath the epidermis, a vascularized dermis provides structural and nutritive support for the overlying matrix of epidermal cells. Although composed of a collagen and elastic tissue network, with interwoven blood vessels, lymphatics, and nerves, the bulk of dermis consists of mucopolysaccharide gel.

In normal skin, two additional cell types, fibroblasts and mast cells, complete the resident cell populations. By contrast, in pathologic conditions such as acute inflammation, the functional properties of resident cells may change substantially and one may observe in skin biopsy specimens the infiltration into dermis of a variety of blood-borne leukocytes. In fact, the cellular composition of cutaneous infiltrates frequently reflects specific disease entities and the characteristics of such infiltrates may be used to aid in the diagnosis of several skin diseases.

Inflammation

The common denominator in hypersensitivity reactions, as well as other chemical sensitivities in skin, is a sequence of tissue changes that fall under the heading of inflammation. Because of the rather uniform composition of skin, inflammation is relatively independent of body location, and exhibits four cardinal signs or symptoms: redness (erythema), swelling (edema), warmth, and pain. Dermatitis or skin inflammation is a reaction to chemical, biological, or physical irritants, sensitizers, or photosensitizers. Critical to the development of dermatitis is the migration of blood-borne proinflammatory cells (leukocytes), which leave the peripheral vascular space and move into an affected tissue. Since different types of cells have been shown to be capable of migration, the composition of the infiltrate will reflect the type of chemoattractive substance liberated from skin, which process is obviously dependent upon an intact vascular supply.

Allergic Contact Dermatitis and Irritant Contact Dermatitis

Important distinctions must be made in the terms commonly used to describe the pathogenesis of the two distinct types of inflammatory reactions that follow the application of injurious chemicals to skin. Irritant contact dermatitis reactions result from direct toxic effect, while allergic reactions result from an immunologically mediated hypersensitivity reaction (Table 1). Unfortunately, specialists in the four

Table 1. Contact Irritant and Contact Allergic Dermatitis in Humans

	Irritant	Allergic
Differences		
Affected individuals:	All	Only those sensitized
Onset:	First exposure	Requires prior sensitization
Chemical concentration:	High	Low
Prototype:	Sodium dodecyl sulfate (surfactant)	5-Pentadecylcatechol (poison ivy hapten)
Similarities		
Extent of disease:	Limited to areas of exposure	
Clinical features:	Heat, swelling (edema), redness (erythema), pain, blisters	

disciplines that frequently evaluate or treat such conditions (dermatologists, ophthalmologists, immunologists, and toxicologists) occasionally employ contradictory terms. This problem cannot be ignored because in recent years scientists in each of these various disciplines have had to communicate with each other with increasing frequency.

Histopathologic Observations in Contact Dermatitis

The first obvious inflammatory event in contact dermatitis is generally seen in the epidermis, where the death of epidermal cells (necrosis) and the formation of small blisters becomes prominent (spongiosis and vesicle formation). At the same time, in underlying dermis one can observe dilatation of blood vessels, the accumulation of fluid (edema), and an infiltration of acute inflammatory cells. This is reflected clinically as redness and swelling. In severe dermatitis, the separation of necrotic epidermal cells may lead to a complete disruption of the epidermal barrier, allowing escaping serum to produce a superficial crust.

Immunology

Immunology is the scientific discipline that studies the regulation and expression of acquired inflammatory reactivities. An immunologic reaction is specific, acquired, transferable (with cells or antibodies), and exhibits memory. This last feature, memory, is an important characterizing feature of contact allergic dermatitis because it reveals that prior exposure to a chemical is required before the inflammatory response may be elicited (expressed). The mechanisms by which immune responses occur have been divided into two relatively independent, but interactive processes: those mediated by cells and those mediated by certain antibodies found in serum.

A critical concept in understanding an immune response is that it divides into two phases or limbs; in the afferent limb it is acquired (learning), and in the efferent limb it is elicited (expression). With respect to UV effects on immunologic processes, both limbs may be disturbed, but the majority of studies indicate that UV's effect is more important during acquistion than during expression. Since an established immune response persists for extended periods of time (it exhibits memory), an effect of UV (even though the UV is a remote event) will also persist for that extended period of time. On the other hand, a UV effect on elicitation lasts only as long as that reaction does, and seems not to affect the long-term memory.

Photoimmunology

Considerable scientific attention has been paid during the last decade to a capacity of electromagnetic radiation, primarily within the ultraviolet region, to alter certain immunologic reactions that occur in skin. The development of this discipline, called "photoimmunology," has ridden the wave of interest in the immunologic processes that occur normally in skin, and this in turn has demonstrated skin to be an organ of substantial immunologic consequence.[12]

Skin biologists have recognized with great excitement that skin, and particularly the epidermis, plays an important immunologic role that complements its previously recognized function as a barrier. An important part of immunologic function is vested in a minor population of cells that reside permanently within skin: it is now recognized that epidermal Langerhans cells (LC) play a pivotal role in the induction of certain immune processes in skin.[13]

Distinctions Between Eye and Skin

The eye may be conceived of as a specialized extension of brain tissue, forming a direct photoreceptive window with the environment. As its first element of protection, the eye is covered with a specially differentiated ectodermal structure that resembles epidermis, the most superficial portion of skin. Although many structural and functional homologies exist between corneal epithelium and epidermis, their disparate anatomical locations give rise to quite different microbiological and chemical susceptibilities.

Contact Hypersensitivity

Perhaps the most direct photoimmunologic observations have concerned the capacity of UV to distort antigen processing during the induction of contact hypersensitivity (CH). CH is the animal model for human allergic contact dermatitis: the delayed-type hypersensitivity reactions in which a reactive chemical reaches the immune apparatus after topical application.[14]

On first contact, the hapten (for example, dinitrochlorobenzene, methylparaben, or pentadecylcatechol) passes through the stratum corneum barrier and then into viable epidermis and the dermis, where it binds chemically (usually with a covalent bond) to intrinsic structural proteins and cells. Of particular relevance is binding of the haptens to epidermal LC, which are strategically located where they are postulated to play the pivotal process of antigen presentation. Specifically, several *in vitro* tests of LC function have demonstrated that LC are able to play the necessary step of antigen presentation to T lymphocytes, which, in association with other humoral signals, are then induced to proliferate.[15, 16] This proliferative response leads to the generation of a large number of duplicate cellular copies from the original responding lymphocyte, all of which then have the capacity to recognize the original hapten. This recognition and proliferative process involves many cells, and thus takes several days to weeks to complete. It means that large numbers of sensitized cells are acquired over time. Once established, these specifically sensitized T lymphocytes and their progeny will live for a long time, and when the same hapten is encountered they will still be available to migrate into skin where they initiate inflammation. This delayed acquisition of immunologic reactivity and persistence of that response (or memory) should not be confused with a delayed-type hypersensitivity (DTH) reaction, which is delayed only because it takes time for appropriate numbers of cells to migrate into skin. Delayed refers to the 12–48-hr lag between application of hapten and the onset of inflammatory response in previously sensitized persons.

Obviously, the cellular events that occur during immunization for contact hypersensitivity are considerably more complicated than this brief description indicates. First, an "immune response" is not a single process, but represents a large family of responses, some of which tend to enhance the inflammatory process and some of which tend to diminish its effect.[17] In the context of photoimmunology, the most important are those that tend to compete with or diminish the inflammatory effector response, since they appear to be less sensitive to UV.

Antigen Presentation

An important event in the immunologic recognition and destruction of foreign materials is the processing of antigens by specialized populations of cells, called antigen-presenting cells (APC).[18] The proliferation of antigen-specific T lymphocytes requires a cell–cell interaction between lymphocytes and APCs, which depends upon genetically determined cell surface proteins (the Ia antigens). The APC selects and then induces the proper T cells to proliferate, leading to the generation of a population of "sensitized" T lymphocytes.[19] These lymphocytes are later able to mediate the destructive inflammatory response. The proliferation of antigen-specific T lymphocytes is also dependent upon the presence of one or more soluble protein factors, which enhance that process substantially.

An important issue in interpreting the studies descibed below is that the generation of such effector T lymphocytes is not the only immunologic response that results from immunization. In fact, a family of responses occurs, some of which tend to suppress or down-regulate the effector response.[17] The development of such down-regulatory processes also requires antigen presentation, but by a different set of cells. It would appear in this context that such antigen-presenting cells and the processes they initiate, leading to suppression, are more resistant to UV radiation.

Ultraviolet Radiation Distorts the Induction of Contact Hypersensitivity

Two important observations initiated this work: first, that relatively low doses of UV in spectrum B (UVB) would selectively perturb Langerhans cells, as judged by cell-surface ATPase activity, and, second, that this radiation would simultaneously prevent the optimal induction of CH to dinitrofluorobenzene (DNFB).[20] When skin of the abdominal wall was exposed to modest dosages of UVB radiation (100 J/m^2), the morphology of LC was changed substantially, with only minimal changes in other epidermal cells. In parallel with this change, the normal immunizing protocol for DNFB through similarly irradiated skin failed to induce the expected hypersensitivity response. Moreover, a direct correlation was observed between the number of ATPase-positive LC and the capacity of irradiated skin to sustain the induction of hypersensitivity during the time when normal-appearing ATPase-positive LC returned to the radiation-treated skin.

To determine whether competing suppressor responses had persisted in the face of the irradiation, mice that had received their first exposure to DNFB through irradiated skin were painted at a later date with DNFB through unirradiated back skin. It was observed that such mice were unresponsive to this attempt at conventional immunization, and that this unresponsiveness was limited to DNFB. Moreover, the ability to produce a specifically unresponsive state for DNFB occurred only when DNFB was painted on irradiated skin. That is, irradiation of abdominal wall skin did not alter the capacity of back skin to sustain the induction of contact hypersensitivity. This "local" effect of UVB was related directly to the presence of perturbed Langerhans cells, which suggests that their failure to function in the required step of antigen presentation was the reason.[21]

Subsequent experiments showed that mice rendered unresponsive by these relatively low dosages of UVB followed by hapten painting had developed antigen-specific suppressor T lymphocytes that could be transferred adoptively to immunogenically naive recipient mice.[22] The most likely explanation for this observation is that UV delivered *in vivo* impaired the antigen-presenting potential of resident LC, and that in their absence other hapten-derived cells or cell products deliver a tolerogenic signal to the recipient. Suppressor lymphocytes that develop, then, by virtue of their "memory," prevent subsequent attempts to induce the normal effector (destructive) response against that hapten.

Other investigators have demonstrated a similar relationship between UVB-induced alterations in Langerhans cells and the UVB-induced suppression of contact hypersensitivity. Sauder[23] observed that combined UVA and UVB irradiation, delivered to the cells *in vitro* prior to derivitization with trinitrobenzene sulfonate (TNBS), would prevent optimal sensitization of that hapten. Moreover, animals immunized with similarly treated cells were suppressed significantly in their subsequent ability to develop a contact hypersensitivity response with a painting of "normal" skin 6 days later. They also observed that this suppression could be adoptively transferred in a similar fashion to naive recipients with T lymphocytes. Stingl[24] observed that primed epidermal cells would, in a genetically restricted way, induce the proliferation of T lymphocytes. However, when the epidermal cells were irradiated with UVB, either before or after priming, this author observed a dose-dependent inhibition of the proliferative response. These studies add substantially to those demonstrating the capacity of UVB to interrupt antigen presentation by epidermal Langerhans cells and they give direct insight into the earlier *in vivo* studies using similar amounts and qualities of UVB. The most obvious conclusion is that sensitization through UV-irradiated sites results in an alteration in antigen presentation by LC and that this failure is responsible for the capacity of suppression to overdrive.

Phototoxicity

Photosensitivity is a general term used to describe an adverse inflammatory reaction to light, which may be phototoxic or photoallergic in nature[10, 11] (see Table 2). A phototoxic reaction appears as a sunburn on exposed areas and is independent of the immune system. Molecules capable of producing phototoxicity generally have molecular weights less than 500 daltons and possess highly resonant structures capable of absorbing energy from ultraviolet and visible light.[11]

Theoretically, phototoxic reactions will occur in anyone whose skin is exposed to enough radiant energy of the appropriate wavelengths and in whom sufficient numbers of photosensitizing molecules are available. With few exceptions, nearly all phototoxic reactions are produced by UVA. By contrast, normal sunburn reactions are attributable largely to radiation from the UVB portion of the spectrum.[10]

Phototoxic reactions may be photodynamic or nonphotodynamic depending on their dependence on or independence from oxygen, respectively. Radiation of the appropriate wavelength excites a photosensitizer from ground level to a short-lived singlet state, ^1S. In the most efficient photodynamic sensitizers, the ^1S singlet state undergoes conversion to the more energy-rich triplet state, ^3S. Molecules in the triplet state have a much longer half-life and greater opportunity to react with substrates such as protein, DNA, RNA, and membranes.[25–28]

In photodynamic reactions, the triplet state of the photosensitizer reacts by one of two distinct mechanisms, termed type I and type II. In a type I mechanism, the ^3S triplet state of the photosensitizer is reduced by an electron donated from a substrate in the environment. The photosensitizer or substrate must be in close proximity for type I to occur. In the type II mechanism, the ^3S state of the photosensitizer reacts with oxygen, forming a singlet oxygen, which is capable of causing damage to cell components through direct oxidation. Type II reactions are believed to be the most common mechanism for generating the singlet oxygen. The singlet oxygen molecule

Table 2. Mechanisms of UV-Dependent and UV-Independent Chemical Sensitivities in Skin

	Mechanism	
	Toxic	Immunologic
Contact dermatitis (UV-independent)	Contact irritant dermatitis (sodium dodecyl sulfate)	Contact allergic dermatitis (dinitrochlorobenzene)
Photosensitivity (UV-dependent)	Phototoxicity (psoralen)	Photoallergy (tetrachlorosalicylanilide)

generates its destructive action through the photooxidative loss of histidine, methionine, tryptophan, tyrosine, and cysteine in proteins; the photoxidation of guanine base in DNA; and formation of hydroperoxides with unsaturated lipids.[25, 27] Mediators of inflammation in photodynamic reactions include prostaglandins, histamines, kinins, lysosomal enzymes, and peroxides. Examples of materials that cause photodynamic reactions include such dyes as acridine orange, phenothiazines, porphyrins, and anthracene.[25]

Nonphotodynamic reactions occur independently of oxygen and are characterized by the cross-linking of layer molecules. The chemicals most representative of non-photodynamic reactions are the furocoumarins and psoralens. On a molecular level, psoralens intercalate into DNA by weak chemical interactions. Upon UVA exposure, covalent bonds form between adenine or thymine and psoralen. Upon receiving another photon of energy, the psoralen then cross-links with a second pyrimidine base on the adjacent strand of DNA. The result is a cross-linked DNA in which the individual strands fail to separate *in vivo* during DNA replication, or *in vitro* when standard laboratory techniques are used.[25, 27] Furocoumarins also react by a photodynamic process to produce erythema as observed in psoralen photosensitivity of human skin.[11, 25, 29]

Photoallergy

Photoallergy is an acquired altered reactivity dependent on antigen-, antibody-, or cell-mediated hypersensitivity.[11, 29] Photoallergy exhibits two clinical forms: urticaria (hives) and dermatitis.

Solar urticaria describes a group of uncommon conditions in which, after brief exposure to light, multiple urticarial wheals, with and without flares, occur all within several minutes of exposure. Typically, intense itching usually precedes the appearance of lesions, and this reaction will begin to subside within an hour after onset of symptoms. This condition has been associated with degranulation of dermal mast cells at the site of UV exposure, elevated numbers of eosinophils, and the release of neutrophil chemotactic factors and histamine into venous blood near the reaction site.[29, 30] Varieties of solar urticaria have been classified by electromagnetic action spectra and by the ability to transfer this reactivity with patients' serum into normal subjects.[31, 32] Harber has classified solar urticaria, Types I and IV, as having an allergic pathogenesis based on serum factors.[33] Although the nature of such factors has not been identified, it has been speculated that reaginic antibody (immunoglobulin E) forms in response to an abnormal photoproduct. These antibodies bind to mast cells and subsequent exposure to UV initiates an antibody–antigen reaction at the mast cell surface, resulting in degranulation and the release of inflammatory mediators.[30]

Photoallergic contact dermatitis (photo-ACD) resembles allergic contact dermatitis and usually appears as acute dermatitis, occurring primarily on, but not limited to, light-exposed areas of skin. Epidermal spongiosis and vesicles, although not common, may accompany this reaction. One characteristic feature of photo-ACD is a dense perivascular round cell infiltrate in the dermis.[10, 11, 29] This is an important histological feature that can be used to distinguish photo-ACD from phototoxic reactions. In photo-ACD, the role of ultraviolet light is to excite the photosensitizer and not skin. Photo-ACD is not induced if skin is radiated prior to the application of photosensitizer. In addition, the action spectrum for inducing and eliciting photo-ACD reactions closely corresponds with the energy spectrum for the photoallergen as opposed to structures forming skin.[10, 34] Although photosensitizers absorb energy over a wide energy spectra, predominantly the UVA range is associated with photo-ACD.

Two pathways have been proposed to explain formation of photoproducts. In the first pathway, a photoallergen in its excited state reacts with the body's proteins to form an antigen. In this case, the photoproduct becomes the antigen. Musk ambrette and tetrachlorosalicylamide photoproducts are believed to be formed in this fashion. In the second pathway, the excited state of the photoallegen is converted into a simple contact allergen that then binds to proteins. The photoproduct is the contact allergen. The photoproduct of sulfanilamide is thought to be formed by this second pathway.[34]

In both pathways, the hapten carrier is processed by APC, leading to the proliferation of T lymphocytes. Upon subsequent contact with that hapten, those T cells release soluble mediators that increase vascular permeability, recruit additional inflammatory cells, and ultimately produce acute dermatitis. Photo-ACD, like allergic contact dermatitis, imparts long-lived sensitivity to an antigen. To prevent this condition completely, patients must avoid the photosensitizer, and, in some cases, structurally related chemicals as well.[11, 29]

A small proportion of individuals with photo-ACD develop a sensitivity to light itself, presumably in the absence of a photosensitizer. The term "persistent light reactor" is used to describe this phenomenon. Although this reactivity can persist for less than a year, in its most severe form, it may last indefinitely.[11, 29] Mechanisms for persistent light reactivity are unknown, but light reactivity appears to originate from photo-ACD. Willis and Kligman have proposed that minute quantities of the photoallergen are retained in the dermis of persistent light reactors for indefinite periods of time.[10, 35] Other investigators have raised doubts about this explanation since, in some persistent light reactors, the action spectrum for elicitation extends from 290 nm through UVA to visible light, consequently exceeding the typical action spectrum of photoallergens.[10] Three materials associated with persistent light reactivity are halogenated salicylanilides, 6-methylcoumarin, and musk ambrette.[11, 30, 36, 37]

Epstein has categorized exogenous photosensitizers capable of causing photoallergy into two classes, topical and systemic. Topical photosensitizers are chemicals that produce localized reactions following application to skin. Systemic photosensitizers are generally drugs that arrive at the skin through the vasculature as systemically incorporated molecules when administered orally or parenterally.[11, 29] Lists of chemicals and drugs capable of causing photo-ACD may be found in several sources.[6, 26, 30, 34, 38]

Clinical Observations

In 1939, Stephan Epstein first differentiated photoallergy from phototoxicity for the drug sulfanilamide, leading him to postulate that for at least one form of sulfanilamide photosensitivity, immunologic reactivity was required.[39-41] By 1970, several topical photosensitizers had been recognized to have caused an epidemic of photoallergic reactions in the United States and Europe. Wilkinson reported 53 cases of photo-ACD to 3, 3′, 4′, 5-tetrachlorosalicylanide (TSCA),[42] an antimicrobial agent incorporated into soaps to reduce bacterial colonization. In only 1 year, 1960–1961, it was responsible[29, 42] for 10,000 cases of photo-ACD in England. Photo-ACD has also been induced by several of the phenolic compounds introduced to replace TCSA, including bithionol, hexachlorophene dichlorophene, the carbanilides, Fenticlor (bis[2-hyroxy-5-chlorophenyl] sulfide), Jadit (a mixture of buclosamide and salicyclic acid), and chloro-2-phenylphenol.[29] Since 1968, the incidence of photo-ACD has declined following the removal of the most potent photosensitizers from consumer products and at present, photo-ACD presents in only a very small proportion of the patients seen regularly by practicing dermatologists.[10, 29]

Within the last 5 years, two perfume raw materials, 6-methylcoumarin and musk ambrette, have been shown to produce photo-ACD. The synthetic compound, 6-methylcoumarin is structurally related to the furocoumarins. This perfume raw material was used as a fragrance in a variety of cosmetics and toiletries, with concentrations ranging from 0.001 to 0.4%. Several years ago, an epidemic outbreak of photodermatitis was attributed to a sunscreen containing 6-methylcoumarin.[43, 44] Because of the many consumer complaints about this product and the results of an investigative study conducted by Kaidbey and Kligman showing 6-methylcoumarin to be a potent human photoallergen, the Food and Drug Administration recalled all suntan products containing this ingredient.[41, 45] Currently, the International Fragrance Research Association (IFRA) recommends that 6-methylcoumarin and 7-methylcoumarin not be used as fragrance ingredients.[43, 46]

Musk ambrette is a nitromusk compound used in fragrances, soaps, detergents, creams, lotions, and dentifrices and is used in perfumes in concentrations from 1 to 15%. In 1978, Larsen reported the first case of photo-ACD in a patient who used an aftershave lotion that contained musk ambrette.[43] Two years later, Giovinazzo reported two additional patients with photo-ACD attributed to musk ambrette. Over a period of years no other cause was established in these cases despite extensive testing and hospitalization, and photopatch testing identified musk ambrette as a photoallergen. Despite this, both patients remained "persistent light reactors" after removal of all known sources of musk ambrette.[47] IFRA now recommends that musk ambrette not be used in concentrations exceeding 4% in perfumes and colognes. They also recommend that musk ambrette not be used at a concentration exceeding 0.1% in fragrances intended for use in aftershave products.[43, 46]

Photo-ACD Is Cell-Mediated

Evidence that photo-ACD is a cell-mediated process comes from several sources. Dermatologists have maintained that the clinical appearance, histologic characteristics, and diagnostic patch test results of patients with photo-ACD were virtually indistinguishable from those of patients with allergic contact dermatitis.[11] Herman and Sams provided additional indirect evidence through studies suggesting that antibodies do not play an essential role in photo-ACD reactions. Using microOuchterlony and immunofluorescence techniques, they were unable to detect antibodies (IgG, IgM, or IgA) to 3,5 dibromosalicylamide (3,5-DBS) in patients with photo-ACD. *In vitro* studies[29] demonstrated that immunoglobulin fractions of these patients would not bind to cutaneous tissue pretreated with 3,5-DBS. Jung also showed that the lymphocytes from patients photosensitized to Jadit could be activated in a lymphocyte transformation test with a light-induced Jadit–albumin complex but not with Jadit alone.[48, 49]

Laboratory Animal Studies

The selection of an animal model for the study of ACD and photo-ACD was based on classic work of Landsteiner and Chase, who demonstrated that low-molecular-weight haptens produce clinically obvious allergic contact dermatitis in guinea pigs. Most importantly, they observed that ACD could be conferred on immunologically naive guinea pigs by the passive transfer of mononuclear cells from sensitized animals.[50, 51] This seminal observation became the cornerstone of photo-ACD research and, over many years, outbred albino guinea pigs have become the most commonly used animal in photo-ACD research.[52] Guinea pigs are inexpensive, easy to handle and, most importantly, develop a clinically obvious response to contact sensitizers. As in humans, sensitized guinea pigs responded to topically applied sensitizers with both erythema and edema. By contrast, sensitized mice, when exposed to contact allergens, develop only an edematous response with little or no erythema.[53–55]

In working with sulfanilamide, Schwarz and Speck first reported in 1957 the successful induction of experimental photo-ACD in guinea pigs.[56] During the next three decades, other investigators experienced variable success inducing photo-ACD to a wide variety of photosensitizers. Photo-ACD would readily be induced with photoallegens such as TCSA, tribromosalicylamide (TBS), and chlorpromazine. By contrast, other agents such as musk ambrette and 6-methylcoumarin, although potent human photosensitizers, were ineffective in inducing photoallergy in guinea pigs, when conventional protocols were used. This prompted investigators to include other manipulations in their immunization regimens. Agents used to enhance the responses of guinea pigs included combinations of UVA and UVB radiation, sodium lauryl sulfate, cellophane tape stripping, skin abrasion with a high-speed nylon brush, and Freund's complete adjuvant.[57–62] With the exception of Freund's complete adjuvant, each of these techniques was designed to compromise the stratum corneum barrier and thus enhance cutaneous penetration of photosensitizers.

Harber and Baer, using a modification of the technique of Landsteiner and Chase, passively transferred photo-ACD to TCSA from sensitized animals to nonsensitized animals with peritoneal exudate cells.[28] Their studies demonstrated that photo-ACD is a cell-mediated response similar to ACD, requiring UVA for photoproduct formation.

A major recent development has been the use of inbred mice in photo-ACD research. As a predictive model, mice have become a useful and inexpensive alternative to guinea pigs, providing a high degree of correlation with human responses to several photosensitizers.[29, 52] Inbred mice, because of their genetic homogeneity and well-characterized immunologic reactivities, have served well as a model for characterizing the factors that regulate immune responses.[63, 64]

Maguire and Kaidbey reported that inbred mice could develop photo-ACD to a variety of photosensitizers. Photo-ACD was induced by applying photosensitizers to clipped flank skin and then irradiating the application sites with combined UVA and UVB radiation. For weak photoallergens, the adjunct administration of immunopotentiators was required to produce an obvious photo-ACD response. Immunopotentiators consisted of either an intradermal injection of *Corynebacterium parum* at the induction site after skin painting or pretreatment of mice with systemic cyclophosphamide. Seven days later, mice were challenged by painting the photosensitizer on the ears of test and control animals and irradiating both groups with UVA. Ear swelling responses were measured 24 and 48 hours later. With this technique, Maguire and Kaidbey induced photo-ACD in several inbred mice strains CBA/J, Balb/c, C57B1/6, and DBA/2 mice to photosensitizers such as TCSA, chlorpromazine, 6-methylcoumarin, sulfanilamide, and sodium omadine.[65, 66]

Takigawa and Miyachi have produced photo-ACD to TCSA in Balb/c, DBA/2, C3H/He, C57B1/6, B10.A and A/J mice. Their protocol, an adaptation of Claman et al.'s procedure for inducing ACD in mice, consisted of painting a photoallergen on the abdomen, followed by irradiation of the site with UVA on 2 consecutive days. Mice were challenged with the photoallergen by ear painting and irradiation, with ear thickness measured 24 hr later. They, like other investigators, observed that ear swelling peaked at 24 hr following the challenge dose and was near to normal values by 48 hr. They also observed that sensitivity was detectable as early as 3 days after immunization but typically peaked between 5 and 7 days. In contrast to guinea pigs and humans, they observed that photo-ACD to TCSA was limited to BALB/c mice. Moreover, attempts to elicit cross-reactivities with tribromosalicylanilide or bithionol in mice photosensitized to TCSA were unsuccessful.[67]

Takigawa and Miyachi explained the cellular nature of photosensitivity in mice. Photo-ACD to TCSA could be transferred with lymph node cells to naive recipients genetically identical at the major histocompatibility complex, whereas attempts between histoincompatible strains were unsuccessful. Moreover, treatment of lymph node cells with anti-Thy 1.2 monoclonal antibody and complement abrogated the transfer, demonstrating that T lymphocytes were required for photo-ACD. Attempts to transfer photo-ACD to TCSA with immune sera or polymorphonuclear exudate cells were unsuccessful. These studies have demonstrated photo-ACD to TSCA in inbred mice to be a highly specific delayed contact hypersensitivity reaction, requiring T cells, and restricted by genetic products of the major histocompatibility complex.[67]

Within the last 3 years, investigators have reexamined cyclophosphamide for its capacity to modulate the induction of photo-ACD in BALB/c mice. Granstein confirmed that cyclophosphamide will enhance photo-ACD to TSCA in mice, observing that, in its absence, both diminished responses and antigen-specific suppressor T cells may be found.[68, 69] Cyclophosphamide has been known to prevent selectively the generation of suppressor cell activity in delayed contact hypersensitivity reactions, a fact that also accounts for Maguire and Kaidbey's success with cyclophosphamide pretreatment.[65, 66, 68, 69]

Granstein observed that UVB is not required to induce photo-ACD to TCSA in BALB/c mice. UVA radiation during the induction phase was found to be sufficient and the addition of UVB did not increase photo-ACD responses above those observed with UVA radiation alone.[68, 70] This observation contrasts with other studies in which the induction of photo-ACD to TCSA in guinea pigs was enhanced by added UV-B exposure,[71] as well as with the report by Kaidbey that both UVA and UVB were essential for inducing photo-ACD in humans.[38]

Granstein reported that relatively high doses of UVB delivered to a site remote from the induction area before application of the TCSA and UVA radiation would inhibit responses to TCSA. In contrast with the earlier observation that similar doses of UVB radiation would suppress responses to DNCB through the generation of suppressor cells, no suppressor activity could be detected in the spleens of these photosensitized mice, presumably because the cyclophosphamide inhibited their generation.[68–70] Takigawa also reported that preirradiation of the photosensitizing site with UVB 2 days prior to induction depletes the mice of ATPase-positive Langerhans cells and renders mice unresponsive at the time of challenge with TCSA.[72] They observed that the spleens of these mice did contain antigen-specific suppressor T cells that blocked the afferent limb or induction phase of photo-ACD, and they established through cell transfer studies that a cyclophosphamide-sensitive T cell was necessary for the suppressor cell to exert its effects.[73]

This section has "highlighted" some of the recent animal research in photoallergic contact dermatitis, always with an eye toward clinical relevance. In this regard, Maguire and Kaidbey have made mice refractory to the induction of photo-ACD by inducing them intraveneously with photohaptenized cells (spleen cells and photoallergen incubated together and irradiated with UVA light). Work of this sort may eventually lead to tolerizing desensitizing protocols for photo-ACD and for contact hypersensitizing as well.[66]

Variables That Influence the Induction of Photo-ACD

Harber and Shalita have suggested that the induction of photo-ACD is dependent upon the clinical state of skin, body region, and timing. The quantity and application site, capacity of a photosensitizer to penetrate skin, pH, solubility conditions at the site of exposure, the quantity and wavelengths of radiation employed, ambient temperature, humidity, the thickness of the stratum corneum, melanin content, and the integrity of the immune system all play a role.[6] Kaidbey and Kligman reported that the photosensitizing capacity of photoallergens is strongly influenced by the choice of vehicle. In general, emulsion-type creams were superior to petrolatum, while polyethylene glycol was uniformly poor, tending to inhibit the activity of test materials. However, they have stressed that there are no reliable guidelines for forecasting the effect of a base on a photosensitizer.[74]

Predicting Photo-ACD to Products

Most predictive tests for photo-ACD have their origin in basic investigative procedures, which, in turn, were designed primarily to induce photo-ACD in animals for the investigation of immune responses to photoallergens. Because of the epidemic proportions of photo-ACD in the 1960s and 1970s, experimental procedures for inducing photo-ACD in guinea pigs were modified into predictive tests for screening new products and raw materials. Harber, using a modification of the technique of Vinson and Borselli, reported the successful induction of photo-ACD to TCSA.[57] While this method, which combined UVA and UVB exposure, became the standard for inducing photo-ACD, it was eventually found to be inadequate to produce photo-ACD against musk ambrette and 6-methylcoumarin.

Working with Hartley guinea pigs, Harber and Deleo[52] then used pretreatment with Freund's complete adjuvant to increase the incidence of photosensitization against these two sensitizers. To induce photo-ACD, nuchal areas of 10–20 test animals were shaved and depilated. Areas of 6–8 cm^2 were defined by injection of 0.1 ml of Freund's complete adjuvant into four corners. The test materials were then applied to each area under an occlusive patch, as defined by the injection sites. After patch removal, sites were exposed to UVA and, with the exception of adjuvant injection, this induction procedure was repeated four times in 2 weeks. Concurrent with this treatment, negative control animals received identical treatment, except that the sensitizer was not included.

Prior to challenge, an irritation screen was conducted on the backs of 10 animals, using three concentrations of each test material, applied in duplicate on opposite sides of the back. One side was covered with an opaque material to test for irritant contact dermatitis while the other side was irradiated with UVA to test for phototoxicity. Sites were read at 24, 48, and 72 hr following application of test materials.

Two weeks after induction, sensitized and control guinea pigs were challenged with nonirritating concentrations of test materials applied in duplicate. One side received

UVA exposure while the other was marked with an opaque material. Each animal was assessed for reactions at 24 and 48 hr. When test reactions exceed control reactions, the material used was defined as a photosensitizer. When similar reactions are observed on both sides, the test material was defined as a contact sensitizer. Validation of this technique, like other photo-ACD predictive tests, consisted of testing high concentrations of photoallergens in a simple vehicle system such as alcohol, acetone, petrolatum, and mineral oil.

By contrast, toxicologists evaluate formulations containing a multitude of raw materials in complex emulsion systems. Even with adequate controls, problems can arise in this latter circumstance to make data interpretation difficult, if not impossible. Over the last 18 months, our laboratory has used the above protocol to predict the photoallergic potential of several fragrance formulations. Concentrations of fragrances in these products ranged from 4 to 20% weight by volume in alcohol or anionic surfactant systems. The decision to conduct a photo-ACD testing on formulations was made after considering the likelihood that purchasers may apply products before sunlight exposure. Our collective experience with these products has led us to develop some suggestions for using this test.

When testing potential photosensitizers with irritant properties, an irritation screen should be conducted on guinea pigs pretreated with Freund's complete adjuvant. This additional step eliminates false-positive reactions. Freund's complete adjuvant may shift the threshold of skin irritation downward for many irritants. Using nonirritant considerations of test materials during challenge, as determined from an irritation screen with animals not pretreated with Freund's complete adjuvant, results in large amounts of erythema on test and control animals. This recommendation complements the observation of Klecak and Magnusson concerning the use of Freund's complete adjuvant in the guinea pig maximization test.[75] Irritating concentrations can be avoided during challenge by following the above recommendation. When it is not possible to conduct an irritation screen on animals pretreated with Freund's complete adjuvant, toxicologists can avoid selecting an irritating dose by reducing the selected concentration from the irritation screen by a factor of 10%. If one compares scores in test and control groups, this dose will usually differentiate sensitization and photosensitization from skin irritation.

We also recommend that the toxicologist assess the ACD potential of products before starting any photo-ACD tests. If a product is positive for ACD in a modified Buehler test or by split adjuvant technique, then it will be difficult to assess that product's photo-ACD potential. Products containing contact sensitizers will produce positive responses on both flanks of test guinea pigs in a photo-ACD test, preventing one from assessing its photoallergic potential. Similarly, results from a positive ACD reaction in a predictive photo-ACD test should not be relied upon exclusively to conclude that the product is unsafe. In this circumstance, judgment concerning the safety of a product should be based on the collective results of delayed contact hypersensitivity tests (such as the Buehler test) or split adjuvant test, human repeated-insult patch test, provocative use studies, and experience with similar formulations.

Recently, two new protocols for predicting photo-ACD in guinea pigs have been reported. Buehler has developed an induction regimen in which guinea pigs are treated by an occluded patch for 4 hours a day, three times a week for 3 weeks. A specially modified Hill Top™ chamber containing the test materials is applied to the clipped and depilated sites. Depilation appears to be an essential step in each application prior to radiation, since the hair stubble interferes with effective irradiation. After each treatment, test sites are irradiated for 2 hr with 12 black light lamps emitting UVA or, in the case of 6-methylcoumarin, 11 black lights and a single sunlamp for UVB emission. UVB is included to produce erythema. After it was determined that UVA's only exposure was ineffective with 6-methylcoumarin 10–14 days after the last induction treatment, the treatment and control groups are challenged by a single 4-hr occlusive patch applied to a naive depilated site. Two hours after the removal of patches, sites are irradiated for 2 hr. These sites are then scored using a graded scale designed for the modified Buehler test for ACD. According to the investigators, this method is sensitive enough to detect photo-ACD to musk ambrette and 6-methylcoumarin, and can differentiate phototoxicity from photoallergy.[76, 77]

Guillot[78] reports a new method that assesses the phototoxic and photoallergic potentials of chemical in the same guinea pig. Test substances are applied under occlusion to guinea pigs pretreated with Freund's complete adjuvant. Fluorescent lamps emitting UVA and UVB are used to irradiate test sites. After the challenge, animals are scored for erythema and edema and a histologic examination of a lesion, when present, is performed. Based on the percentage of photosensitized animals, a grade of I through V is assigned to reflect weak, mild, moderate, strong, and extreme photosensitizing potentials. In their system, 6-methylcoumarin and sulfanilamide received grades of IV; tribromosalicylamide, a grade of I; bithional, TCSA, and musk ambrette, grade II; and promethazine, a grade of III (2.0% challenge concentration) and IV (5.0% challenge concentration).

Regardless of the animal test systems and protocols used to evaluate the photo-ACD potential of products, toxicologists may have second thoughts about conducting human testing. Predictive ACD tests in humans are designed to maximize the potential to induce an immunologic-based reaction to unknown chemicals. Consequently, these tests should be approached cautiously only after the product has successfully passed predictive ACD and photo-ACD tests in animals. Although this test is a relevant approach to predicting photo-ACD in humans, it nevertheless has the potential to produce persistent light reactions. Because of the serious and debilitating nature of this condition, alternatives such as testing the product in two animal species such as the guinea pig and mouse should be exhausted before a predictive human photo-ACD test is conducted.

Diagnostic Tests in Humans

The ability of UV to produce a spectrum of disorders in humans has led clinical investigators to develop clinical procedures to characterize their causes since many phototoxic and photoallergic reactions are indistinguishable.

Photopatch Testing

The easiest and most definitive test to employ is the photopatch test. For this procedure, several commercial companies have prepared many of the common sensitizers in relatively pure form and at appropriate concentrations. Although most are not both photocontact sensitizers and contact sensitizers, it is important in all studies to employ negative unirradiated control sites to rule out this possibility.

Photoallergy

To test for photoallergic reactions, 0.01-cc samples of a series of sensitizers are placed in duplicate on a small metal plate which is then fastened to the skin surface with tape. Because of this close approximation, some of each chemical will penetrate the skin surface similarly to that seen in a simple contact dermatitis. After 24 or 48 hr, the appliance is removed and half of the duplicate samples are examined to rule out irritant and allergic reactions, and they are then covered rapidly with light barrier. The other half is cleansed lightly of excess sensitizer and then exposed to an appropriate amount of radiation from a UVA source or to a broad-spectrum UV source with window glass filters. These exposed sites are then also covered with an opaque barrier. Twenty-four hours later, all sites are examined for evidence of acute dermatitis and graded according to severity: redness alone; redness and edema; or redness, edema, and blister formation. A positive response in a UVA-treated site alone represents a photoallergic reaction, whereas a positive response in both irradiated and unirradiated sites represents contact allergic dermatitis at least, and, on occasion, both simultaneously. Although it is easy to describe this procedure, one must recognize that there are many chances for false-positive and false-negative errors. One important aspect that makes the performance of photopatch testing relatively easy is the large difference between the amount of UVA necessary to activate photoallergy and the amount of UVA that is directly phototoxic.

Photosensitivity

Photosensitivity testing is best accomplished with a monochromatic light source that can be used to assess the wavelength of sensitivity. This is particularly useful since *in vivo* the action spectrum should reflect the absorption spectrum of the drug in question. For photosensitivity testing to systemically administered drugs, however, patients cannot serve as their own negative control, and one must rely on comparisons with other similarly tanned control subjects. When an individual is severely sensitive, this is not difficult, but when sensitivity is only several times that of natural sensitivity, particularly in the UVB region, differentiation from normal deviations is difficult.

References

1. J. A. Parrish, R. R. Anderson, F. Urbach, and D. Pitts, Skin aging and carcinogenesis due to ultraviolet radiation, In *UV-A: Biological Effects of Ultraviolet Radiation with Emphasis on Human Responses to Longwave Ultraviolet*, Plenum Press, New York, 1978.
2. A. L. Lenhninger, Photosynthetic electron transport and phosphorylation, In *Biochemistry, The Molecular Basis of Cell Structure and Function*, Worth Publishers, New York, 1970.
3. J. W. Longworth, Light, colors and origins of spectroscopy, In *The Science of Photomedicine*, J. D. Regan and J. A. Parrish, Eds., Plenum Press, New York, 1982.
4. J. A. Parrish, Photobiology and immunology, In *The Effect of Ultraviolet Radiation on the Immune System*, J. A. Parrish, Ed., Johnson and Johnson Baby Products Company, Skillman, New Jersey, 1983.
5. B. Ranby and J. F. Rabels, Fundamental principles of photochemistry, In *Photodegradation, Photooxidation and Photostabilization of Polymers: Principles and Applications*, John Wiley, New York, 1975.
6. L. C. Harber, A. R. Shalita, and R. B. Armstrong, Immunologically mediated contact photosensitivity in guinea pigs, In *Dermatotoxicology*, 2nd Edition, F. N. Marzulli and H. I. Maibach, Eds., Hemisphere Publishing, Washington, 1983.
7. J. A. Parrish, R. R. Anderson, F. Urbach, and D. Pitts, The spectrum of electromagnetic radiation, In *UV-A Biological Effects of Ultraviolet Radiation with Emphasis on Human Responses to Longwave Ultraviolet*, Plenum Press, New York, 1978.
8. J. A. Parrish, R. R. Anderson, F. Urbach, and D. Pitts, Uses of UV-A involving exposure of humans, In *UV-A Biological Effects of Ultraviolet Radiation with Emphasis on Human Responses to Longwave Ultraviolet*, Plenum Press, New York, 1978.
9. J. A. Parrish, R. R. Anderson, F. Urbach, and D. Pitts, Immediate and short-term biologic effects of ultraviolet radiation on normal skin, In *UV-A Biological Effects of Ultraviolet Radiation with Emphasis on Human Responses to Longwave Ultraviolet*, Plenum Press, New York, 1978.
10. J. A. Parrish, R. R. Anderson, F. Urbach, and D. Pitts, Adverse cutaneous reactions to UV-A, In *UV-A Biological Effects of Ultraviolet Radiation with Emphasis on Human Responses to Longwave Ultraviolet*, Plenum Press, New York, 1978.
11. J. H. Epstein, Phototoxicity and photoallergy in man, *J. Amer. Acad. Derm.* *8*(2):141, 1983.
12. J. W. Streilein, Skin-associated lymphoid tissue (salt): Origins and functions, *J. Invest. Derm.* *80*:125, 1983.
13. P. R. Bergstresser and J. W. Streilein, Langerhans cells participation in contact hypersensitivity, In *Progress in Diseases of the Skin*, Volume 2, R. Fleischmajir, Ed., Grune and Stratton, New York, 1984.
14. P. R. Bergstresser, Immunologic mechanisms of contact hypersensitivity, In *Dermatologic Clinics: Contact Dermatitis*, F. J. Storrs, Ed., W. B. Saunders, Philadelphia, 1984.
15. G. Stingl, S. I. Katz, L. Clement, I. Green, and E. M. Shevach, Immunological function of Ia-bearing epidermal Langerhans cells, *J. Immunol.* *121*:2005, 1978.
16. J. P. Jakway and E. M. Shevach, Stimulation of UV-treated, antigen-pulsed macrophages: Evidence for antigen presentation and interleukin 1 production, *Cell Immunol.* *80*:151, 1983.
17. G. L. Asherson and M. Zembala, Suppression of contact sensitivity by T cells in the mouse: I. Demonstration that suppressor cells act on the effector stage of contact sensitivity and their induction following *in vitro* exposure to antigen, *Proc. R. Soc. Biol.* *187*:329, 1974.
18. D. W. Thomas, G. Farni, E. M. Shevach, and I. Green, The role of the macrophage on the stimulation cell in contact sensitivity, *J. Immunol.* *118*:1677, 1977.
19. R. N. Germain, Accessory cell stimulation of T cell proliferation requires an active antigen processing Ia-restricted antigen presentation and a separate nonspecific second signal, *J. Immunol.* *127*: 1964, 1981.

20. G. B. Toews, P. R. Bergstresser, and J. W. Streilein, Epidermal Langerhans cell density determines whether contact hypersensitivity or unresponsiveness follows skin painting with DNFB, *J. Immunol. 124*:445, 1980.
21. P. R. Bergstresser and J. W. Streilein, Ultraviolet radiation produce selected immune incompetence, *J. Invest. Derm. 81*:85, 1983.
22. C. A. Elmets, P. R. Bergstresser, R. E. Tigelaar, P. J. Wood, and J. W. Streilein, Analysis of the mechanisms of unresponsiveness produced by haptens painted on skin exposed to low dose ultraviolet radiation, *J. Exp. Med. 158*:781, 1983.
23. D. N. Sauder, K. Tamaki, A. N. Moshell, H. Fujiwara, and S. I. Katz, Introduction of tolerance to topically applied TNCB using TNP-conjugated ultraviolet light-irradiated epidermal cells, *J. Immunol. 127*:261, 1981.
24. G. Stingl, L. A. Gazze-Stingl, W. Aberer, and K. Wolff, Antigen presentation by murine epidermal Langerhans cells and its alteration by UVB light, *J. Immunol. 127*:1707, 1981.
25. A. Korhauser, W. Wamer, and A. Giles, Jr., Light-induced dermal toxicity: Effects on the cellular and molecular level, In *Dermatotoxicology*, 2nd Edition, F. N. Marzulli and H. I. Maibach, Eds., Hemisphere, Washington, 1983.
26. L. C. Harber, I. E. Kochevar, and A. R. Shalita, Mechanisms of photosensitization to drugs in humans, In *The Science of Photomedicine*, J. D. Regan and J. A. Parrish, Eds., Plenum Press, New York, 1982.
27. J. D. Spikes, Photodynamic reactions in photomedicine, In *The Science of Photomedicine*, J. D. Regan and J. A. Parrish, Eds., Plenum Press, New York, 1982.
28. L. C. Harber and R. L. Baer, Pathogenic mechanisms of drug-induced photosensitivity, *J. Invest. Derm. 58*:327, 1972.
29. J. H. Epstein, Photoallergy and photoimmunology, In *Dermatologic Immunology and Allergy*, J. Stone, Ed., C. V. Mosby, St. Louis, 1985.
30. W. L. Morison, Photoimmunology and humans, In *The Effects of Ultraviolet Radiation on the Immune System*, J. A. Parrish, Ed., Johnson and Johnson Baby Products Company, Skillman, New Jersey, 1983.
31. H. Ive, J. Lloyd, and I. A. Magnus, Action spectra in idiopathic solar urticaria, *Brit. J. Derm. 77*:229, 1965.
32. W. M. Sams, Jr., Solar urticaria: Studies of the active serum factor, *J. Allergy Clin. Immunol. 45*:295, 1970.
33. L. C. Harber, R. M. Holloway, V. R. Wheatley, and R. L. Baer, Immunologic and biophysical studies in solar urticaria, *J. Invest. Derm. 44*:439, 1963.
34. I. E. Kochevar, Photobiologic mechanisms in photoimmunology: Membrane alterations and photobiology, In *The Effects of Ultraviolet Radiation on the Immune System*, J. A. Parrish, Ed., Johnson and Johnson Baby Products Company, Skillman, New Jersey, 1983.
35. I. Willis and A. M. Kligman, The mechanism of the persistent light reactor, *J. Invest. Derm. 51*:385, 1968.
36. J. H. Epstein, K. D. Wuepper, and H. I. Maibach, Photocontact dermatitis to halogenated salicylanides and related compounds, *Arch. Derm. 97*:236, 1968.
37. S. Brown, P. R. Lane, and I. Magnus, Skin photosensitivity from fluorescent lighting, *Brit. J. Derm. 81*:420, 1969.
38. K. Kaidbey, Topical photosensitizers, In *Safety and Efficacy of Topical Drugs and Cosmetics*, A. M. Kligman and J. J. Leyden, Eds., Grune and Stratton, New York, 1982.
39. S. Epstein, Photoallergy and primary phototoxicity to sulfanilamide, *J. Invest. Derm. 2*:243, 1939.
40. S. Epstein, Photoallergy and primary photosensitivity to sulfanilamide, *Dermatologica 83*:63, 1941.
41. S. Epstein, Photoallergy versus phototoxicity, In *Dermatology Due to Environmental and Physical Factors*, R. B. Rees, Ed., Charles C. Thomas, Springfield, Illinois, 1962.
42. D. S. Wilkinson, Photodermatitis due to tetrachlorosalicylamide, *Brit. J. Derm. 73*:123, 1969.
43. W. G. Larsen, Perfume dermatitis, *Amer. Acad. Derm. 12*(1):1, 1978.

44. K. H. Kaidbey and A. M. Kligman, Photocontact allergy to 6-methylcoumarin, *Contact Derm. 4*:277, 1978.
45. H. J. Eiermann, Regulatory issues concerning AETT and 6-ME, *Contact Derm. 6*:120, 1980.
46. Monograph on Fragrance Raw Materials, Research Institute for Fragrance Materials, Inc., Englewood Cliffs, New Jersey.
47. V. J. Giovanazzo, L. C. Harber, R. B. Armstrong, and I. E. Kochevar, Photoallergic contact dermatitis to musk ambrette: Clinical reports of two patients with persistent light reactor patterns, *Amer. J. Acad. Derm. 3*(4):384, 1980.
48. E. G. Jung, U. Dummler, and H. Immich, Photoallergie durch 4-chlor-2-hydroxy-benzoesaure-n-butylamid, Lichtbiologische untersuchungen zur antigenbildung, *Arch. Klin. Exp. Derm. 232*:403, 1968.
49. E. G. Jung, J. Hornke, and P. Hajdu, Photoallergie durch 4-chlor-2-hydroxy-benzoesaure-n-butylamid. II Photochemische untersuchungen, *Arch. Klin. Exp. Derm. 233*:287, 1968.
50. K. Landsteiner and M. W. Chase, Experiments on transfer of cutaneous sensitivity to simple compounds, *Proc. Soc. Exp. Biol. Med. 49*:688, 1942.
51. K. Landsteiner, *The Specificity of Serological Reactions*, Harvard University Press, Cambridge, Mass., 1945.
52. L. C. Harber and V. A. DeLeo, Guinea pigs and mice as effective models for the prediction of immunologically mediated contact photosensitivity, In *Models in Dermatology, Dermatopharmacology and Toxicology*, H. I. Maibach and N. J. Lowe, Eds., Karger, New York, 1985.
53. A. J. Crowle and C. M. Crowle, Contact sensitivity in mice, *J. Allergy 32*:302, 1961.
54. H. Möller, Allergic contact dermatitis in the mouse ear, *Acta Dermatovener 61*:1, 1981.
55. M. A. Vadas, J. F. A. P. Miller, J. Gamble, and A. Whitelaw, A radiosotope method to measure delayed type hypersensitivity in the mouse. I. Studies in sensitized and normal mice, *Int. Arch. Allergy Appl. Immunol. 49*:670, 1975.
56. K. Schwarz and M. Speck, Experimentelle untersuchungen zur frage der photoallergie der sulfonamide, *Dermatologica 111*:232, 1957.
57. L. C. Harber, S. E. Targovnik, and R. W. Baer, Contact photosensitivity patterns to halogenated salicylanilides in man and guinea pigs, *Arch. Derm. 96*:646, 1967.
58. D. J. Cripps and T. Enta, Absorption and action spectra studies on bithionol and halogenated salicylanilides photosensitivity, *Brit. J. Derm. 82*:230, 1970.
59. T. Horio, The induction of photocontact sensitivity in guinea pigs without UVB radiation, *J. Invest. Derm. 67*:591, 1976.
60. W. P. Jordan, Jr., The guinea pig as a model for predicting photoallergic contact dermatitis, *Contact Derm. 8*:109, 1982.
61. T. H. Mauer, E. G. Weirich, and R. Hess, Predictive animal testing for photocontact allergenicity, *Brit. J. Derm. 103*:593, 1980.
62. Ichikawa, R. B. Armstrong, and L. C. Harber, Photoallergic contact dermatitis in guinea pigs: Improved induction technique using Freund's complete adjuvant, *J. Invest. Derm. 76*:498, 1981.
63. J. Klein, *Biology of the Mouse. Histocompatibility-2 Complex*, Springer-Verlag, New York, 1975.
64. D. H. Katz and B. Benaceraf, *The Role of Products of the Histocompatibility Gene Complex in Immune Responses*, Academic Press, New York, 1976.
65. H. C. Maguire and K. Kaidbey, Experimental photoallergic contact dermatitis: A mouse model, *J. Invest. Derm. 73*(3):147, 1982.
66. H. C. Maguire and K. Kaidbey, Studies in experimental photoallergy, In *The Effects of Ultraviolet Radiation on the Immune System*, J. A. Parrish, Ed., Johnson and Johnson Baby Products Company, Skillman, New Jersey, 1983.
67. M. Takigawa and Y. Miyachi, Mechanisms of contact photosensitivity in mice: I. T. cell regulation of contact photosensitivity to tetrachlorosalicylanilide under the genetic restrictions of the major histocompatibility complex, *J. Invest. Derm. 78*:108, 1982.

68. R. D. Granstein, Effects of UVB radiation on photocontact hypersensitivity in the murine model, In *The Effects of Ultraviolet Radiation on the Immune System*, J. A. Parrish, Ed., Johnson and Johnson Baby Products Company, Skillman, New Jersey, 1983.
69. R. D. Granstein, W. L. Morison, and M. L. Kripke, The role of suppressor cells in the induction of murine photoallergic contact dermatitis and in its suppression by prior UVB irradiation, *J. Immunol. 130*(5):2099, 1983.
70. R. D. Granstein, W. L. Morison, and M. L. Kripke, The role of UVB radiation in the induction and elicitation of photocontact hypersensitivity to TCSA in the mouse, *J. Invest. Derm. 80*:158, 1983.
71. P. S. Herman and W. M. Sams, *Soap Photodermatitis, Photosensitivity to Halogenated Salicylanilides*, Charles C. Thomas, Springfield, Illinois, 1972.
72. Y. Miyachi and M. Takigawa, Mechanisms of contact photosensitivity in mice: II. Langerhans cells are required for successful induction of contact photosensitivity to TCSA, *J. Invest. Derm. 78*:363, 1982.
73. M. Takigawa, Y. Miyachi, K. Toda, and A. Yoshioka, Mechanisms of contact photosensitivity in mice. IV. Antigen-specific suppressor T cells induced by preirradiation of photosensitizing site of UVB, *J. Immunol. 132*(3):1124, 1984.
74. K. H. Kaidbey and A. M. Kligman, Topical photosensitizers. Influence of vehicles on penetration, *Arch. Derm. 110*:868, 1974.
75. G. Klecak, Identification of contact allergens: Predictive tests in animals, In *Dermatotoxicology*, 2nd Edition, F. N. Marzulli and H. I. Maibach, Eds., Hemisphere, Washington, 1983.
76. E. A. Newmann, and R. D. Parker, A device for irradiating guinea pig skin with ultraviolet light, *Food Chem. Toxicol. 23*(7), 1985.
77. E. V. Buehler, E. A. Newmann, and R. D. Parker, The use of occlusive patch to evaluate the photosensitive properties of chemicals in guinea pigs, *Food Chem. Toxicol. 23*(7), 1985.
78. J. P. Guillot, J. F. Gonnet, J. F. Loquerie, M. C. Martini, P. Convert, and J. Cotte, A new method for the assessment of phototoxic and photoallergic potentials by topical applications in the albino guinea pigs, *J. Toxicol. Cut. Ocular Toxicol. 4*(2):117, 1985.

4

PHOTOCARCINOGENESIS

P. DONALD FORBES, Ph.D.
RONALD E. DAVIES, Ph.D.
FREDERICK URBACH, M.D.
CURTIS COLE, M.S.
Center for Photobiology
Skin and Cancer Hospital
Temple University Health Sciences Center
Philadelphia, Pennsylvania

Abstract

The role of ultraviolet radiation (UVR) in the induction of human skin cancer is clear and of major importance. In extensive experiments with UVR on animal (mouse) skin, skin cancer has been produced by repeated exposure to wavelengths between 250 and 320 nm, with the 290–320-nm range being most effective.

For the appropriate interpretation of experimental results, factors of importance include choice of experimental animal (haired vs. hairless, age, pigmentation, genetic background), spectral composition of the light source, dose rate, and dose delivery. Spectral component additivity and dose–time reciprocity are among the major areas that require further investigation.

The interrelationships between radiation, chemicals, and skin pose special problems. Possible light–chemical interactions that may affect photocarcinogenesis include photochemical alteration of the chemical, ability of the chemical to attach to DNA with formation of strong bonds after exposure to UVR, and enhancement of photocarcinogenesis by a process similar to "tumor promotion."

Photocarcinogenesis appears to be one manifestation of a larger photobiological process involving both local and systemic responses to UVR.

Introduction

The high-energy, short-wavelength portion of the solar electromagnetic spectrum (less than 320 nm) is potentially very detrimental to living cells and tissues. A low concentration of ozone formed in the stratosphere absorbs most of the photons of ultraviolet radiation (UVR) and thus prevents most of them from reaching the earth. However, even in the presence of this ozone layer, which varies in thickness at different latitudes and different seasons, a biologically significant amount of UVR reaches the surface of the earth.

The major effects on humans of radiation in the UVB range (280–320 nm) are on the skin and eyes. Acute effects consist of "sunburn," an inflammatory response of the tissues that may be no more than mild redness or slight stinging of the eyes, or may develop into the equivalent of second-degree (blistering) burns. The acute effects of single overdoses of UVB are transient, heal without scarring, and in the skin lead to adaptive changes of skin thickening and pigmentation, which afford some degree of protection. The only established positive (beneficial) effect of UVB in humans is the production of vitamin D precursors in the skin, which are absorbed into the bloodstream and prevent rickets, a serious vitamin deficiency disease. Most work has been done on the harmful effects of UVB and relatively little attention has been given to its possible beneficial effects, however. Repeated UVB exposure, prolonged over years, can result in chronic degenerative changes in skin, characterized by skin "aging" and the development of premalignant and malignant skin lesions.

Within a relatively short period of time, there have appeared in the scientific literature three concepts that stimulated interest in photocarcinogenesis. One was the concern that human activities could adversely affect the ability of the earth's atmosphere to remove harmful solar ultraviolet radiation by optical filtration;[1,2] a second was the realization that photocarcinogenesis could be enhanced greatly by a variety of chemicals;[3-5] a third was the discovery that UVR could affect photocarcinogenesis via the immune system.[6,7] In each case, the concept enriched the field of photobiology while contributing to an understanding of immediately relevant phenomena. This review surveys some of the factors influencing the response of UV-irradiated subjects.

The case for associating nonmelanoma skin cancer (NMSC) with UVR rests on a line of circumstantial evidence whose cogent points are as follows[8-10]:

It is clearly established that NMSC occurs most frequently on the head, neck, arms, and hands, parts of the body habitually exposed to sunlight, and where histologic changes of chronic UVR damage are most severe.

Pigmented races, who sunburn much less readily than people with white skin, develop much less NMSC and when it does occur, it most frequently affects areas not exposed to sunlight.

Among whites there appears to be a much greater incidence of NMSC in those who spend more time outdoors than those who work predominantly indoors.

NMSC is more common in white-skinned people living in areas where insolation is greater.

Genetic diseases resulting in greater sensitivity of skin to the effect of solar UVR are associated with markedly increased and premature skin cancer development (albinism, xeroderma pigmentosum).

Skin cancer can be produced readily on the skin of mice and rats with repeated doses of UVB; that is, the same spectral range that produces erythema solare (sunburn) in human skin.

The material in this chapter focuses on animal studies, but it is useful to review the morphologic basis for accepting experimental models as relevant to human problems. UVR is thought to induce a variety of tumor types in human skin, with squamous cell and basal cell carcinomas predominating.[11] Solar keratoses, carcinoma *in situ*, and keratoacanthomata probably are related developmentally or anatomically to squamous cell carcinoma; UVR apparently induces few sarcomas in human skin. The relationship of UVR to melanoma is still conjectural;[12] there remains a paradox in the strong relationship of malignant melanomas to exposure–latitude gradients (albeit in limited geographic areas), but lack of concentration of melanomas on demonstrably exposed sites. Even after corrections for improved diagnostic techniques, the incidence and death rate from malignant melanoma are rising rapidly in many developed societies.[13] If increased exposure to sunlight is responsible for this rise, the effective wavelengths remain a mystery; the available data do not any more readily implicate the UVR spectrum than the visible or infrared. A combination of factors is very likely involved.[12-14]

UVR-exposed animals produce a variety of epithelially derived tumors, including carcinomas and benign epitheliomas; they can yield sarcomas under some circumstances. UVR-irradiated mice develop essentially no distinguishable basal cell carcinomas; they have not yet provided a useful model for malignant melanoma in skin. Investigators have taken a variety of approaches to recognizing, describing, measuring, and recording skin tumors.[6, 15-22] Data have been based on tumor size, external morphology, location, behavior, and histopathologic findings.

Animal Models

Although skin cancers are induced in nature by chronic exposure to ultraviolet radiation in various animals (the eye of cattle,[23] skin of pigs, hairless dogs), the only practical animal models are mice and rats. For reasons of efficacy and economy, mice have almost uniformly been used as experimental animals in photocarcinogenesis studies.

The Influence of the Animal Model on Tumor Response

Anatomical Differences

The keratin of hair is a very effective scattering and absorbing medium, and consequently animals with hair are susceptible to UV-induced tumors only on the relatively naked ears, tail, and snout. Even when the hair is clipped off close to the skin surface, the hair stubble provides some protection. When the hair is chemically depilated, leaving no stubble, the sensitivity of haired mice increases significantly (Forbes, unpublished observations). Whether the skin is electrically clipped or chemically depilated, UV-irradiated haired mice tend to produce tumors that are morphologically distinct from those of their nonhaired counterparts.[6, 15, 17, 19, 24]

"Hairless" and "rhino" mice have phenotypes whose controlling genes are alleles.[25] Both geneotypes produce a juvenile coat of hair that is lost at the end of its growth phase; the animals do not grow a normal coat of hair subsequently. Rhino mice develop a thicker, folded skin, with a more hyperplastic epidermis and adherent stratum corneum than that of the hairless mice. Both rhino and hairless mice respond to UV radiation by producing carcinomas, but the hairless mice are significantly more susceptible. The mixed mutant allelic phenotype is distinguishable from either a hairless or rhino parent, with an intermediate skin condition. We have derived homozygote hairless (*hr/hr*) and intermediate hairless/rhino (hr/hr^{rh}) mice with common genetic backgrounds and have exposed them to simulated sunlight.[26] The dose–response curves of these two lines of animals are similar and parallel, with a significant offset in absolute sensitivity. This is most easily seen by comparing lines representing the median latent periods for selected daily doses of UV radiation.

Melanin pigment has long been associated with protection of human skin from damage by UV radiation. The pigmented ears of a black-haired strain of mice were shown to be less susceptible to UV-radiation-induced cancer than ears of an albino strain.[27] The back skin of animals with pigmented extremities is also less susceptible than the skin of albino mice, even though the skin produces very little visible pigment.[28] Other heritable factors may contribute to the differences in response, as are discussed later in this chapter.

Differences in Tumor Response in the Absence of Anatomical Dissimilarity

We have found that strains of albino hairless mice are significantly different in their susceptibility of tumors induced by simulated sunlight.[28] Grube[29] previously reported that two strains responded differently to the carcinogenic stimulus of broad-spectrum UVA radiation plus 8-methoxypsoralen. In contrast, these strains had similar dose relationships for such acute symptoms as erythema and the formation of epidermal cell thymine dimer cross-links. Hypotheses for their dissimilar susceptibility to UV-radiation-induced cancer include subtle differences in skin structure, epidermal cell kinetics, vascular responses, and immunologic capability.

Effects of Age

The ear skin of haired mice[30] and the dorsal skin of hairless mice[31] become somewhat less susceptible to the formation of UV-radiation-induced tumors as age increases. Ebbesen and Kripke[32] have found a similar change with age in the skin of clipped haired mice exposed to fluorescent sunlamps. Black[22] has included age as a variable in a part of studies on antioxidants. In this case, the age of the animals did not appear to influence susceptibility to UV-radiation-induced carcinogenesis.

Indirect Effects

One test system clearly demonstrates that tumor growth is under immunologic control. Ultraviolet irradiation of C_3H haired mice apparently induces appearance of cells that inhibit the animals' ability to reject UV-radiation-induced tumors.[6, 7, 33] Through numerous elaborate experiments, the involvement of T-lymphocyte suppressor cells has been demonstrated. The system apparently can react differently to UV radiation alone or to photosensitization (psoralen plus UVA radiation).[34]

De Gruijl and Van der Leun[18] have demonstrated a systemically mediated effect of UV radiation on photocarcinogenesis. Preexposing one side of hairless mice to UV radiation makes the other side more susceptible to subsequent tumor induction by UV radiation. Whether this systemic effect is immunologically mediated remains to be determined.

Identification and Measurement of the Carcinogen

The influences of action spectrum, of the UVR dose delivered, and of the means of UVR delivery are conceptually distinct, even though they are all implicit in a given experiment. Each has been tested as a variable in one or more studies relevant to this review.

Action Spectra

Sunlight is made up of photons with a broad spectrum of energies, potencies for photobiological interactions, and likelihood of influencing other photobiological reactions. An "action spectrum" is a description of the relative effectivenes of photons at each wavelength (or energy level), in producing a defined response. Action spectra have been determined for a number of photobiological responses. The most familiar method utilizes a UVR source and a "monochromator" beam including wavelength ranges of no more than a few nanometers. This method has technical limitations (such as the surface area that can be exposed), and can be used only with great difficulty in studies on the production of tumors. Some monochromator-based data are available for tumor development on the ear of the mouse,[35] and similarities to the human erythema action spectra were noted. This study also illustrates the instrumental and biological difficulty in acquiring such data using nominally monochromatic radiation. Additional studies would be required to determine whether the individual wavebands were strictly additive in their effects.

Luckiesh employed another approach to the determination of human skin response (erythema and tanning) spectra, incorporating a broad-band light source with a series of short-wavelength cutoff filters.[36] The action spectrum was estimated from a matrix of simultaneous equations. We have used similar methods to study carcinogenic effectiveness in hairless mice, employing a series of simulated solar UVR spectra.[37]

The action spectrum for photocarcinogenesis cannot, of course, be determined in humans. The alternative endpoint most extensively studied in humans is the acute "sunburn" effect, particularly at the level of minimal response (perceptible erythema). A number of studies have established the action spectrum for this response; within limits imposed by the facts that the end point is somewhat subjective and the exact characterization of light sources is difficult, these studies are in rather good qualitative and quantitative agreement.[38] The action spectra thus derived have been used as estimates of the spectral weighting function for carcinogenesis in many of the mathematical simulations presently under investigation.

The "minimal erythema" threshold spectrum in humans corresponds closely to acute-response action spectra in hairless mice (principally for the edema response).[39, 40] It has further been shown that the monochromatically derived action spectrum accurately predicts the effectiveness of polychromatic sources for producing the same response (over the fairly limited series of spectral qualities tested). Finally, this action spectrum reasonably predicted the effectiveness of a series of polychromatic spectra for producing carcinogenesis in hairless mice.[41] Thus, in the mouse, extrapolation from monochromatic estimates of an acute-effects action spectrum to production of the chronic effect of a polychromatic source has been validated, lending greater credibility to the use of this extrapolation in humans. Moreover the demonstration of qualitatively and quantitatively similar acute response spectra in the two species strengthens the extrapolation of carcinogenesis studies in mice to the human problem.

Effectiveness of UVA Alone

For a laboratory UV source producing a mixed spectrum (in which at least 2% of the UV energy is UVB), the UVA appears to contribute little to its carcinogenic effectiveness.[41] This is not to say that UVA may be safely ignored; under cetain circumstances, UVA alone is demonstrably effective.[42] With the advent of "tanning bed" fixtures, which might deliver sizable quantities of UV radiation in which the UVB component is less than 2%, the working model described above could underestimate carcinogenic effectiveness. This possibility has been considered[43] and it is clear that we need more information on human exposures and additional data from laboratory studies.

Dose–Response Relationships

A description of dose–response relationships has fundamental importance, first, in predicting the impact of exposure to UVR under specified circumstances, and, second, in attempting to identify the nature of the carcinogenic process. Some of the problems of defining "response" have been described;[17] we will now consider the meaning of "dose." (SI nomenclature[44] substitutes the word "fluence" for "dose;" the former implies somewhat more information than is usually available in skin studies, but the difference is not important in the present context.) The term "dose–response" connotes a quantitative correspondence between delivery of UVR and the production of tumors. The simplest case would be represented by delivering, within a few minuts on 1 day, a variety of doses to groups of mice, using a single UVR source, and (much later) finding a linearizing relationship between dose magnitude and some measure of tumor production. Skin tumors can be produced in mice with a single brief exposure to UVR, but the amount of radiation required to produce even a low incidence of tumors is an ulcerating dose.[45] Most UVR-cancer studies involve protracted or repeated exposures over some appreciable fraction of the animals' lifetime. Thus a "lifetime dose," or at least a dose to a given tumor endpoint, represents a cumulative figure derived from many exposures. Most of the laboratory data on photocarcinogenesis come from studies in which the exposure is repeated from one to seven times per week for several weeks. Blum demonstrated the quantitative nature of tumor response in the ears of chronically irradiated haired mice[16] and we reported a similar study using body skin of hairless mice.[17] The effect of increasing the amount of UVR per exposure is to shorten the time required to produce tumors, and to increase the numbers of tumors produced at a specified time. The time to 50% incidence is inversely proportional to log dose per exposure (provided the tolerance of skin to acute damage is not exceeded). A detailed analysis of UVR-cancer dose–response in mice has been reported recently by de Gruijl.[46]

Dose Delivery

In the absence of information to the contrary, one assumes time–dose reciprocity, that is, that the effect of a stimulus will be proportional to the total dose delivered (or absorbed), where dose is the product of exposure duration and intensity. This relationship often holds quite well for primary photochemical systems, but the time required to deliver the dose becomes a significant factor in biological systems[31, 47] and many cases of "reciprocity failure" have been demonstrated. For example, the carcinogenic effectiveness of a given dose of ionizing radiation is decreased when the dose is spread over a longer delivery period;[48] in contrast, the carcinogenic effectiveness of DMBA on hairless mouse skin is increased by extending the time between topical applications.[49] Data from our animal studies indicate rather consistently that both fractionation and attenuated delivery of UVR increase the carcinogenic effectiveness of a given dose.[17, 31, 50] Changes in radiation quantity and quality within dosing cycles or over the total period of dose delivery remain to be investigated. Deviations from dose–time reciprocity may pose the most formidable problem for extrapolation from laboratory studies to human epidemiology.

Wavelength Interaction in Photocarcinogenesis

The concept of action spectrum or relative efficacy of different wavelengths is well-documented with regard to the UVB region. Application of an action spectrum to predictions of the efficacy of polychromatic sources requires an assumption of additivity, for example, that the effectiveness of a source is the sum of the effectiveness of its spectral components. This assumption is the focus of major disagreements in photobiology.

Investigators have differed on action spectra for photobiological effects of the UVA region.[36] There is therefore little agreement on appropriate weighting(s) to be applied to this wavelength region, and nominally unweighted energy integrals are often presented as the simplest method of describing treatment conditions. Interpretation of studies can be further complicated by inaccurate description of radiation sources or by unrecognized alterations in commercially obtained sources.[51]

The possibility exists that various parts of the spectrum may act in other than an additive fashion. With acute erythema as an endpoint, for example, evidence has been presented to indicate that the wavelengths longer than 320 nm either enhance or inhibit the effectiveness of UVB radiation. Van Weelden[52] reported that effectiveness of an exposure to UVB radiation can be decreased by subsequently reexposing the area to UVA and visible radiation. In contrast, Willis[53] reported that exposure to UVA radiation enhances the effectiveness of UVB radiation, particularly when the exposure to UVA radiation precedes the UVB radiation. Ying[12] has reported that the combined effect can be attributed to additivity of effectiveness by various portions of the spectrum. Willis[54] has also reported that the response to chronic UVA and UVB radiation exposures (epidermal hyperplasia and solar keratosis) shows an interactive effect (greater than additivity).

UVA Enhancement of the Carcinogenic Effect of UVB

Based in part on evidence from acute-effects studies, warnings against human exposures to UV radiation include the statement that UVA can enhance the carcinogenic effect of UVB.[55] The direct experimental evidence for such a statement is, at best, conflicting. Urbach[56] described an analysis of experimental results on carcinogenesis obtained with various light sources, suggesting such an interaction. In subsequent studies Forbes[50] found no evidence that UVA enhanced carcinogenesis by UVB; under certain circumstances UVA even appeared to reduce it. In the meantime, Willis[20] introduced the term "photoaugmentation" in association with a method for rapidly inducing skin cancer in mice. Although they did speed up the process of tumorigenesis, their report is difficult to interpret. Like the first studies examined by Urbach,[56] their exposures to UVB and to the combined UVB + UVA were different in various respects (apart from the amount of UVA), notably in the spectral composition of the UVB, and the irradiance. More recently, Van der Leun and van Weelden[57] exposed one group of mice to UVB and a second group to the same UVB, with UVA given simultaneously. The combination of UVB and UVA proved to be slightly less carcinogenic than UVB alone. These authors called this process photorecovery, the reverse of photoaugmentation. Both terms may be premature, since they tend to suggest explanations for effects whose existence is not clearly established. A similar experiment was reported by Staberg;[58] UVA given simultaneously with UVB appeared to reduce the carcinogenic effect of UVB, although this result was not statistically significant.

The exposures to UVA discussed so far were given either simultaneously with UVB or immediately before or after each of a series of UVB exposures. Several experiments have been performed in which mice were exposed to UVB over a period of time, followed by a period of exposures to UVA. Forbes[50] exposed mice to UVB from fluorescent sunlamps during several weeks, with doses sufficient to result in tumors later in their lives. For half of the mice, this sequence of exposures was followed by exposures to UVA (320 nm) during 40 weeks; the other mice did not receive any further exposures. The mice that received UVB followed by UVA developed markedly fewer tumors than the animals receiving UVB only.

Staberg[59] reported on a similar experiment. During the first period of 3 months, mice received solar-simulated UVB and, during the second period, the light from unfiltered fluorescent UVA lamps (these lamps also emitted some UVB). These investigators concluded that their UVA exposures, although not carcinogenic in themselves, enhanced the carcinogenic effect of the previously administered artificial sunlight. Their data, however, do not exclude the possibility that the increased carcinogenic effect was due to a cumulative action of the carcinogenic stimuli given during the two successive periods. This interpretation is supported by the finding in a subsequent experiment,[58] which was similar but in which most of the UVB was filtered from the UVA lamps, that the observed increase by the UVA exposures was substantially reduced.

In conclusion, the studies pointing to enhancement of UVB carcinogenesis by UVA are not conclusive. Studies in which UVA appears to be the only variable do not confirm the phenomenon, and some show the reverse effect. At least with sunlike radiation, it appears possible to account for photocarcinogenesis based on assumptions of wavelength additivity.

Interaction of Chemicals

Several modes of UVR–chemical interactions have been reviewed by Davies[60] and by Robinson.[61] In many cases, the possibility of interaction between UVR and a given chemical is dependent on the timing of the two treatments. There are examples of several types of interaction: the chemical itself may be irradiated prior to contact with the biological target, or irradiation may occur after the chemical has been reacted with the tissue; in some cases, irradiation may have to be restricted when the chemical is present in the tissue. For a given chemical, the results of these choices could be quite different. For example, psoralens are effective only if intercalated in cellular DNA before exposure to UVR. Chemical binding is temporary, and so effectiveness diminishes as the interval increases. In contrast, the interactive effects of retinoic acid depend on its being protected from direct UVR exposure to avoid photodestruction.

Photomodified Chemical Carcinogenesis

The interrelationships among light, chemicals, and skin carcinogenesis were first noted by Findlay[62] in 1928. Since early studies of chemically induced skin carcinogenesis were performed using coal tar as the inducing agent, attention was drawn to the observation of Lewin that coal tars exhibited photodynamic action.[63] Soon after, the purification of the phenanthrene constituents, 7,12-dimethylbenzanthracene and benzpyrene, showed evidence of photodynamic action in tissue culture, and with each of these the activating wavelengths were found in the UVA range.[64]

Combined UVR and visible radiation and polynuclear aromatic hydrocarbon (PAH) carcinogen treatment of animal skin has produced conflicting results. Acceleration of tumorigenesis had been reported.[62, 65] Clark showed that UVA irradiation, immediately after painting with 20-methylcholanthrene, accelerated papilloma formation without affecting incidence of carcinomas;[66] this effect was, up to a point, dose-related, but prolonged or intense UVR had an inhibiting effect on tumor formation. Stenback[67] found that pretreatment of mouse skin with UVB caused a marked increase of 7,12 dimethylbenzanthracene- (DMBA) induced skin tumors, while irradiation after application of the carcinogen significantly decreased tumor induction. In contrast, others[68-71] found either no effect or a decrease in skin tumor production, although most of these observers reported development of acute skin phototoxicity. In most of these studies radiation delivery was concurrent with or subsequent to chemical treatment, both of which conditions favor both phototoxicity and photochemical degradation of the carcinogen.

Santamaria[72] concluded that benzpyrene is carcinogenic independently of UVR, but the benzpyrene-induced carcinogenicity can be influenced by UVR or visible light irradiation, causing either acceleration *or* inhibition of tumorigenesis. He suggested that acceleration of tumor production was due to a benzpyrene-dependent photodynamic reaction compatible with cellular life, and that inhibition was due to severe photodynamic damage to cells and extracellular components, rather than to alteration of the carcinogen by photooxidation.

Schonberg had already suggested that "light" might destroy externally applied carcinogens: the suggested mechanism was photodimerization, which is known to occur in some anthracene-type compounds.[73]

Potency of DMBA is reduced by UVA in accord with demonstrable photochemical activity of the compound.[74] Either preirradiating DMBA *in vitro*, or treating animals with DMBA and immediately irradiating them with blue light, resulted in many fewer tumors than were observed in animals treated with unexposed DMBA and kept in red light.[70] On the other hand, if animals pretreated with DMBA were exposed to intense UVR, a phototoxic reaction was induced, accompanied by a somewhat increased tumor yield.

The decreased biological activity was shown to be dependent on the photochemical properties of DMBA. Irradiation at 360 nm produced almost pure nonfluorescent transannular peroxide, as reported by Cook and Martin;[75] however, shorter wavelengths of light produced further degradation to multiple products.[76] Each of the five major products found, as well as mixtures of all photoproducts, were shown to have negligible carcinogenic activity.

Photochemical Carcinogenesis

Photoinduced carcinogenesis has been observed following topical application of agents that are phototoxic, but not in themselves carcinogenic. Büngeler reported that subcutaneous injection of eosin, hematoporphyrin, and a tar solution enhanced the ability of light to produce skin tumors.[77] His experiments are difficult to interpret, since the mortality in each of the experimental groups was very high, and occurred long before tumors appeared.

Miescher[78] did not succeed in producing tumors in mice exposed five times weekly to a clinically phototoxic combination of UVA and anthracene. He concluded that there were no reasons for viewing photodynamic action as "a new, specific carcinogenic agent." Heller, in an attempt to reproduce Miescher's results, produced necrotizing phototoxic skin changes in mice using UVA and anthracene. The mice developed neoplasms at the borders of the phototoxically produced skin ulcerations.[79] Heller's conclusion was diametrically opposed to that of Miescher: that photodynamic reactions "should be recognized as a new carcinogenic principle."

The introduction of 8-methoxypsoralen (8-MOP) as a therapeutic agent for certain human skin diseases was followed by reports that it could enhance photocarcinogenesis in mice.[80, 81] It was shown that skin cancer could be induced by the interacting effects of a compound (8-MOP) and irradiation with a particular waveband of light (UVA); neither of these agents is a primary carcinogen under the conditions of the test.

Forbes[82] investigated the relative enhancing effect of two widely recognized photoactive compounds (8-MOP and anthracene) on photocarcinogenesis; neither is a chemical carcinogen. Compared with the vehicle alone, topical treatment with the

8-MOP solution, but not the anthracene solution, markedly enhanced photocarcinogenesis under conditions in which neither material produced acute phototoxicity.

Chemically Enhanced Photocarcinogenesis

Epstein and Roth showed that croton oil would "promote" the growth of tumors initiated by exposing skin to UVR.[83] Epstein reported promotion-like enhancement of photocarcinogenesis in the presence of all-trans retinoic acid (RA).[84] The term "promotion" is somewhat misleading, considering its usual implications in chemical carcinogenesis, because most such studies have involved delivery of an intended minimally carcinogenic dose of radiation in a limited time (typically 3–10 weeks), with the further intent that little or no evidence of carcinogenesis would exist when this dosing was completed. Animals are subsequently exposed to some additional variable (retinoic acid, phorbol ester, benzoyl peroxide, or a different radiation source) to detect "enhancement" of the minimal effective stimulus. Utilizing topical application of RA in methanol, preceded by simulated sunlight irradiation, we also found marked enhancement of skin photocarcinogenesis in hairles mice,[85] in another strain of mice, and in a stock of non-haired rats. We have also examined the effect of RA and tetradecanyl phorbal acetate (TPA) on hairless mice pretreated with either UVR or DMBA. Groups receiving each initiating regimen were treated with vehicle alone, or RA, or TPA for 20 weeks. Both TPA and RA accelerated the appearance of tumors initiated by DMBA. Tumors initiated by UVR were also promoted by both compounds. Virtually all tumors were pedunculated papillomas typical of chemical carcinogenesis but were unlike the squamous cell carcinomas produced by chronic UV irradiation. Thus RA is an effective promoter of cutaneous tumors initiated by either UV or DMBA. Hartmann and Teelman[86] have also reported enhancement of photocarcinogenesis by RA. Verma[87] reported that RA, under certain circumstances, potentiated the formation of skin tumors by DMBA.

Inhibition

The effectiveness of carcinogens in producing skin tumors may be reduced by both chemical and physical means. Mice fed a vitamin-A-supplemented diet developed fewer skin papillomas after a single dose of DMBA than do similar mice fed a diet deficient in vitamin A.[88] It could therefore be anticipated that other retinoids would offer protection against the development of skin cancer. Indeed, such protection has been demonstrated under certain circumstances.[87, 89–92]

Several antioxidants, including BHT and ascorbic acid, have been reported to reduce the effectiveness of UVR in producing skin cancer.[93, 94] This observation is supported by reports that peroxides can promote chemical carcinogenesis.[95, 96]

Melanin pigment in skin provides a natural protection against the damaging effects of UVR;[27] absorption and scattering of UVR photons undoubtedly occur, but melanin may provide the basis of other interactions as well. The free-radical-trapping ability of melanin has been noted,[97] although no direct test of this effect has been reported in photocarcinogenesis studies. Chemicals that reduce the quantity of UVR reaching the target layer in skin can provide protection against photocarcinogenesis.[98, 99]

Practical problems inevitably involve several variables. As an example, let us assume that experimental data support the hypotheses that cosmetic sunscreen preparations prevent at least some UVB radiation from reaching sensitive cells in skin, that (as discussed previously) UV exposures during youth are more carcinogenically effective than those received later in life, and that attenuated exposures are more carcinogenically effective than "pulses" of energy that produce sunburn. Given these assumptions, the strategy of sunscreen use to reduce long-term changes in skin would go beyond merely the avoidance of occasional sunburn.

Other Influences

There are limited data available on a variety of modifying interactions. An increase in susceptibility to UV-induced damage has been reported in the presence of elevated room air temperature, relative humidity, and wind speed.[100] Indoor fluorescent lighting reportedly influences mouse mammary tumor incidence.[101] Ionizing radiation may be synergistic with UVR in producing chronic clinical skin changes,[102, 103] but no recent studies have made a systematic investigation of x-ray effects on photocarcinogenesis. Experimental work may be stimulated by the observation that previous ionizing radiation therapy is one of the variables suspected in the increased susceptibility of patients receiving PUVA phototherapy to carcinoma.[104]

Summary

Photocarcinogenesis, produced by UVB radiation, is the major cause of skin cancer in humans. Experimental studies provide valuable insights into the many variables influencing the photocarcinogenic process. Although tumor production is observed as a localized phenomenon, there is evidence that systemic effects of UVR have important effects on the carcinogenic process.

References

1. National Academy of Sciences, *Biological Impact of Increased Intensities of Solar Ultraviolet Radiation*, Washington, D.C., 1973.
2. National Academy of Science, Report of the Committee on Impacts of Stratospheric Change (CISC), In *Protection Against Depletion of Stratospheric Ozone by Chlorofluorocarbons*, Washington, D.C., 1979.

3. E. A. Emmett, Ultraviolet radiation as a cause of skin tumors, *CRC Crit. Rev. Toxicol.* *2*:211, 1973.
4. H. S. Black and J. T. Chan, Experimental ultraviolet light cacinogenesis, *Photochem. Photobiol.* *26*:183, 1977.
5. J. Epstein, Chemicals and photocarcinogenesis, *Australas. J. Derm.* *18*:57, 1977.
6. M. L. Kripke, J. S. Lofgreen, and J. Beard, In vivo immune responses of mice during carcinogenesis by ultraviolet radiation, *J. Natl. Cancer Inst.* *59*:1227, 1977.
7. R. A. Daynes, M. K. Schmitt, L. K. Roberts, and C. W. Spellman, Phenotypic and physical chracteristics of the lymphoid cells involved in the immunity to synergenic UV-induced tumors, *J. Immunol.* *122*:2458, 1979.
8. F. Urbach, P. D. Forbes, R. E. Davies, and D. Berger, Cutaneous photobiology: Past, present and future, *J. Invest. Derm.* *67*:209, 1976.
9. O. D. Laerum and O. H. Iversen, *The Biology of Skin Cancer (Excluding Melanomas), Report* #15, UICC Technical Report Series, Volume 63, Hans Huber Publishers, Bern, Switzerland, 1981.
10. R. G. Mora and R. Burris, Cancer of the skin in blacks: A review of 128 patients with basal-cell carcinoma, *Cancer* *47*:1436, 1981.
11. C. K. Steigleder, Classification of skin tumors and associated lesions, *Natl. Cancer Inst. Monogr.* *50*:221, 1978.
12. P. Cutchis, On the linkage of solar ultraviolet radiation to skin cancer, Report No. FAA-EQ-78-19, (IDA paper #P-1342) U.S. Dept. of Transportation, Washington, D.C., 1978.
13. J. A. H. Lee, G. R. Petersen, R. G. Stevens, and K. Vesanen, The influence of age, year of birth and date on mortality from malignant melanoma in the populations of England and Wales, Canada and the white populations of the U.S., *Amer. J. Epidemiol.* *110*:734, 1979.
14. J. A. H. Lee and D. Stickland, Malignant melanoma: Social status and outdoor work, *Brit. J. Cancer* *41*:757, 1980.
15. F. Stenback, Life history and histopathology of ultraviolet light induced skin tumors, In *International Conference on Ultraviolet Carcinogenesis*, M. L. Kripke, Ed., U.S. Government Printing Office, Washington, D.C., 1978.
16. H. F. Blum, *Carcinogenesis by Ultraviolet Light*, Princeton University Press, Princeton, N.J., 1959.
17. P. D. Forbes, H. F. Blum, and R. E. Davies, Photocarcinogenesis in hairless mice: Dose–response and the influence of dose-delivery, *Photochem. Photobiol.* *34*:361, 1981.
18. D. D. Grube, R. D. Ley, and R. J. M. Fry, Photosensitizing effects of 8-methoxypsoralen on the skin of hairless mice II. Strain and spectral differences for tumorigenesis, *Photochem. Photobiol.* *26*:269, 1977.
19. J. H. Epstein and W. L. Epstein, A study of tumor types produced by UV light in hairless and hairy mice, *J. Invest. Derm.* *41*:463, 1963.
20. I. Willis, J. Menter, and N. J. Whyte, The rapid induction of cancer in the hairless mouse utilizing the principle of photoaugmentation, *J. Invest. Derm.* *76*:404, 1981.
21. L. H. Kligman and A. M. Kligman, Lack of enhancement of experimental photocarcinogenesis by retinoic acid, In *Retinoids: Advances in Basic Research and Therapy*, C. E. Orfanos, O. Braun-Falco, E. M. Farber, C. Grupper, M. K. Polano, and R. Schuppli, Eds., Springer-Verlag, New York, 1981, p. 411.
22. H. S. Black, V. McCann, and J. I. Thornby, Influence of animal age upon antioxidant modified UV-carcinogenesis, *Photobiochem. Photobiophys.* *4*:107–118, 1982.
23. D. E. Anderson, Cancer eye in cattle, *Mod. Vet. Practice* *51*:43–47, 1970.
24. R. K. Winkelman, E. J. Baldes, and P. E. Zollman, Squamous cell tumors induced in hairless mice with UV light, *J. Invest. Derm.* *34*:131, 1960.
25. S. J. Mann, Varieties of hairless-like mutant mice, *J. Invest. Derm.* *56*:170–173, 1971.
26. P. D. Forbes, Experimental photocarcinogenesis: An overview, *J. Invest. Derm.* *77*:139–143, 1981.
27. H. P. Rusch and C. A. Baumann, Tumor production in mice with ultraviolet radiation, *Amer. J. Cancer* *35*:55–62, 1939.

28. P. D. Forbes, Hairless mice for carcinogenesis studies, Proceedings of the Second NCI/EPA/NIOSH Collaborative Workshop: Progress on Joint Environmental and Occupational Cancer Studies, Sheraton/Potomac, Rockville, Maryland, September 9–11, 1981.
29. D. D. Grube, R. D. Ley, and R. J. M. Fry, Photosensitizing effects of 8-methoxypsoralen on the skin of hairless mice—II. Strain and spectral differences for tumorigenesis, *Photochem. Photobiol.* *25*:169-276, 1977.
30. H. F. Blum, H. G. Grady, and J. S. Kirby-Smith, Relationships between dosage and rate of tumor induction by ultraviolet radiation, *J. Natl. Cancer Inst.* *3*:91-97, 1942.
31. P. D. Forbes, R. E. Davies, and F. Urbach, Aging, environmental influences, and photocarcinogenesis, *J. Invest. Derm.* *73*:131–134, 1979a.
32. P. Ebbesen and M. L. Kripke, Influence of age and anatomic site on ultraviolet carcinogenesis, *J. Natl. Cancer Inst.* *68*:691, 1982.
33. L. K. Roberts, M. Schmitt, and R. A. Daynes, Tumor susceptibility generated in mice treated with subcarcinogenic doses of 8-methoxypsoralen and long-wave ultraviolet light, *J. Invest. Derm.* *73*:306–309, 1979.
34. P. D. Forbes and R. E. Davies, Factors that influence photocarcinogenesis, In *Photoimmunology*, J. A. Parrish, M. L. Kripke, and W. L. Morisin, Eds., Plenum Press, New York, 1983.
35. R. G. Freeman, Data on the action spectrum for ultraviolet carcinogenesis, *J. Natl. Cancer Inst.* *55*:1119, 1975.
36. M. Luckiesh, L. L. Holladay, and A. M. Taylor, Erythemal and tanning effectiveness of UV energy, *Gen. Electric Rev.* *42*:274, 1939.
37. P. D. Forbes, R. E. Davies, F. Urbach, D. Berger, and C. Cole, Simulated stratospheric ozone depletion and increased UV radiation: Effects on photocarcinogenesis in hairless mice, *Cancer Res.* *42*:2796, 1982.
38. D. Berger, F. Urbach, and R. E. Davies, The action spectrum of erythema induced by UV radiation, In *Proceedings of the 13th International Congress of Dermatology*, W. Jadassohn and C. Schirrer, Eds., Springer Verlag, Berlin, 1968, p. 1112.
39. C. Cole, P. D. Forbes, and R. E. Davies, Action spectra for acute cutaneous ultraviolet radiation damage: Man and mouse, *Photochem. Photobiol.* *37*:623, 1982.
40. C. A. Cole, C. Sambuco, R. E. Davies, and P. D. Forbes, Ear swelling in mice as a quantitative response to UVR, Presented at Amer. Soc. for Photobiology, 11th annual meeting, Madison, Wisconsin, *Photochem. Photobiol.* *37*:77s, 1983.
41. C. Cole, C. Sambuco, R. E. Davies, and P. D. Forbes, Skin cancer action spectrum in hairless mice, Presented at Amer. Soc. for Photobiology, 12th annual meeting, Philadelphia, PA, *Photochem. Photobiol.* *39*:13s, 1984.
42. P. D. Forbes, R. E. Davies, and C. A. Cole, Photocarcinogenesis by UVA Alone, Presented at Amer. Soc. for Photobiology, 10th annual meeting, Vancouver, B.C., Canada, 1982.
43. H. van Weelden, F. de Gruijl, and J. C. van der Leun, Carcinogenesis by UV-A; with an attempt to assess the carcinogenic risks by tanning with UV-A and UV-B, In *The Biological Effects of UV-A*, F. Urbach and R. W. Gange, Eds., Praeger, Philadelphia, 1986, in press.
44. D. S. Rupert and R. Latarjet, Toward a nomenclature and dosimetric scheme applicable to all radiations, *Photochem. Photobiol.* *28*:3, 1978.
45. J. Hsu, P. D. Forbes, L. C. Harber, and E. Lakow, Induction of skin tumors in hairless mice by a single exposure to UV radiation, *Photochem. Photobiol.* *21*:185, 1975.
46. F. R. de Gruijl, The dose–response relationship for UV-tumorigenesis, Ph.D. thesis, University of Utrecht, Netherlands, 1982.
47. Y. LeGrand, *An Introduction to Photobiology*, American Elsevier, New York, 1967.
48. R. I. Ullrich and J. B. Storer, Influence of gamma irradiation on the development of neoplastic disease in mice III. Dose-rate effects, *Radiat. Res.* *80*:325, 1979.

49. U. Iversen, O. H. Iversen, and R. Bjerknes, A comparison of the tumourigenic effect of five graded doses of 3-methylcholanthrene applied to the skin of hairless mice at intervals of 3 or 14 days, *Acta Pathol. Microbiol. Scandinav. 73*:502, 1968.
50. P. D. Forbes, R. E. Davies, and F. Urbach, Experimental ultraviolet photocarcinogenesis: Wavelength interactions and time-dose relationships, *Natl. Cancer Inst. Monogr. 50*:31, 1978.
51. P. D. Forbes, R. E. Davies, L. C. D'Aloisio, and C. Cole, Emission spectrum differences in fluorescent blacklight lamps, *Photochem. Photobiol. 24*(6):613, 1976.
52. H. van Weelden, Photorecovery in human skin, In Lasers in Photomedicine and Photobiology (Quantum Electronics Division conference, Florence, Italy 3–6 Sept. 1979). R. Pratesi and C. A. Sceahi, Eds., Proc. European Soc., p. 129-133. Florence, Italy, 1980.
53. I. Willis, A. Kligman, and A. Epstein, Effects of long ultraviolet rays on human skin: Photoprotective or photoaugmentative?, *J. Invest. Derm. 59*:416, 1977.
54. I. Willis, J. Menter, and J. H. Whyte, The rapid induction of cancers in the hairless mouse utilizing the principle of photoagumentation, *J. Invest. Derm. 76*:404, 1981.
55. D. R. Bickers, J. H. Epstein, T. B. Fitzpatrick, L. C. Harber, M. Pathak, and F. Urbach, Risks and benefits from high intensity ultraviolet sources used for cosmetic purposes, *J. Amer. Acad. Derm. 12*:380, 1985.
56. F. Urbach, J. H. Epstein, and P. D. Forbes, Ultraviolet carcinogenesis: Experimental, global and genetic effects, In *Sunlight and Man*, T. B. Fitzpatrick, M. A. Pathak, L. C. Harber, M. Seiji, and A. Kukita, Eds., University of Tokyo Press, Tokyo, 1974, pp. 259–283.
57. J. C. van der Leun and H. van Weelden, Wavelength interactions in reactions of the skin to light, *Photochem. Photobiol. 37*:S81, 1983.
58. B. Staberg, H. C. Wulf, P. Klemp, T. Poulsen, and H. Brodthagen, The carcinogenic effect of UVA irradiation, *J. Invest. Derm. 81*:517–519, 1983.
59. B. Staberg, H. C. Wulf, T. Poulsen, P. Klemp, and H. Brodthagen, Carcinogenic effect of sequential artificial sunlight and UV-A irradiation in hairless mice: Consequences for solarium "therapy," *Arch. Derm. 119*:641–643, 1983.
60. R. E. Davies, Interaction of light and chemicals in carcinogenesis, *Natl. Cancer Inst. Mongr. 50*:45, 1978.
61. H. C. Robinson, The role of synergism in chemical and physical agents as reported in NIOSH criteria documents, M.S. dissertation, University of Cincinnati, Ohio, 1980.
62. G. M. Findlay, Ultraviolet light and skin cancer, *Lancet 2*:1070, 1928.
63. L. Lewin, Uber photodynamische Wirkungen von Inhaltsstoffen des Steinkohlenteer Pech's am Menschen, *Much. Med. Wochenschr. 60*:1529, 1913.
64. M. R. Lewis, The photosensitivity of chick embryo cells growing in media containing certain carcinogenic substances, *Amer. J. Cancer 25*:305, 1935.
65. E. Bingham and P. J. Nord, Cocarcinogenic effects of n-alkanes and ultraviolet light in mice, *J. Natl. Cancer Inst. 58*:1099, 1977.
66. J. H. Clark, The effect of longwave ultraviolet radiation on the development of tumors induced by 20-methylcholanthrene, *Cancer Res. 24*:207, 1964.
67. F. Stenback, Studies on the modifying effect of ultraviolet radiation on chemical skin carcinogenesis, *J. Invest. Derm. 64*:253, 1975.
68. I. Doniach and J. C. Mottram, On the effect of light upon the incidence of tumors in painted mice, *Amer. J. Cancer 39*:234, 1940.
69. J. J. Morton, E. M. Luce-Clausen, and E. B. Mahoney, Visible light and skin tumors induced with benzpyrene in mice, *Cancer Res. 2*:256, 1942.
70. A. C. Kohn-Speyer, Effect of ultraviolet radiation on the incidence of tar cancer in mice, *Lancet 1*:1305, 1929.
71. M. G. Seelig and Z. K. Cooper, Light and tar cancer, *Surg. Gynecol. Obstet. 56*:752, 1933.
72. L. Santamaria, Photodynamic action and carcinogenicity, In *Recent Contributions to Cancer Research in Italy*, P. Bucalossi and U. Veronesi, Eds., Vol. 1, Rome, 1960, p. 107.

73. A. Schonberg, A. Mustafa, M. Z. Barakat, N. Latif, R. Moubasher, and A. Mustafa, Photochemical reactions. Part XIII. (a) Photochemical reactions of ethylenes with penanthraquinone and with 1,2,3-triketones. (b) Dimerisation reactions in sunlight, *J. Chem. Soc.* 2126, 1948.
74. R. E. Davies, H. A. Dodge, and L. H. DeShields, Alteration of the carcinogenic activity of DMBA by light, *Proc. Amer. Assoc. Cancer Res. 13*:14, 1972.
75. J. H. Cook and R. H. Martin, Polycyclic aromatic hydrocarbons, XXIV, *J. Chem. Soc.* 1125, 1940.
76. M. K. Logani, W. A. Austin, and R. E. Davies, Photooxygenation of 7,12-dimethylbenz(a)anthracene, *Tetrahedron Lett. 29*:2467, 1977.
77. W. Büngeler, Uber den Einfluss Photosensibilisierender Substanzen auf die Enstehung von Hautgeschwülsten, *Z. Krebsforsch. 46*:138, 1937.
78. G. Mieschler, Experimentelle Untersuchungen über Krebserzeugung durch Photosensibilisierung, *Schweiz Med. Wochenschr. 23:1082, 1942.*
79. *W. Heller, Experimentelle Untersuchungen* über den Lichtkrebs, *Strahlenther 81*:529, 1950.
80. A. C. Griffin, R. E. Hakim, and J. Knox, The wavelength effect upon erythemal and carcinogenic response to psoralen treated mice, *J. Invest. Derm. 31*:289, 1958.
81. F. Urbach, Modification of ultraviolet carcinogenesis by photoactive agents, *J. Invest. Derm. 32*:372, 1959.
82. P. D. Forbes, R. E. Davies, and F. Urbach, Phototoxicity and photocarcinogenesis: Comparative effects of anthracene and 8-methoxypsoralen in the skin of mice, *Fd. Cosmet. Toxicol. 14*:303, 1975.
83. J. H. Epstein and H. L. Roth, Experimental ultraviolet light carcinogenesis: A study of croton oil promoting effects, *J. Invest. Derm. 50*:387, 1968.
84. J. H. Epstein, Chemicals and photocarcinogenesis, *Australas, J. Derm. 18*:57, 1977.
85. P. D. Forbes, F. Urbach, and R. E. Davies, Enhancement of experimental photocarcinogenesis by topical retinoic acid, *Cancer Letters 7*:85, 1979.
86. H. R. Hartmann and K. Teelman, The influence of topical and oral retinoid treatment of photocarcinogenicity in hairless albino mice, In *Retinoids: Advances in Basic Research and Therapy*, C. E. Orfanos, O. Braun-Falco, E. M. Farber, C. Grupper, M. K. Polano, and R. Schuppli, Eds., Springer-Verlag, New York, 1981, p. 447.
87. A. K. Verma, E. A. Conrad, and R. K. Boutwell, Induction of mouse epidermal ornithine decarboxylase activity and skin tumors by 7,12-dimethylbenz(a)anthracene: Modulation by retinoic acid and 7,8-benzoflavone, *Carcinogenesis 1*:607, 1980.
88. R. E. Davies, Effect of vitamin A on 7,12 DMBA-induced papillomas in rhino mouse skin, *Cancer Res. 27*:237, 1967.
89. D. N. Mulay and F. Urbach, Local therapy of oral leukoplakia, with vitamin A, *Arch. Derm. 78*:637, 1958.
90. W. Bollag, Prophylaxis of chemically induced papillomas and carcinomas of mouse skin by vitamin A acid, *Eur. J. Cancer 8*:689, 1973.
91. M. B. Sporn, N. M. Dunlop, D. L. Newton, and J. M. Smith, Prevention of chemical carcinogenesis by vitamin A and its synthetic analogs, *Fed. proc. 35*:1332, 1976.
92. J. Epstein, Effects of retinoids on ultraviolet-induced carcinogenesis, *J. Invest. Derm. 77*:144, 1981.
93. H. S. Black, Kleinhans, H. T. Hudson, P. M. Sayre, and P. P. Agin, studies on the mechanism of protection by butylated hydroxytoluene to UV-radiation damage, *Photobiochem. Photobiophys, 1*:119, 1980.
94. L. Pauling, R. Willoughby, R. Reynolds, B. E. Bliasizell, and S. Lawson, Incidence of squamous cell carcinoma in hairless mice irradiated with UV light in relation to intake of ascorbic acid (vitamin C) and of D,L- -tocopheryl acetate (vitamin E), Publication #150 from the Linus Pauling Inst. of Sci. and Med., Palo Alto, CA 94306, Third Internat. Symp. on Vitamin C, Sao Paulo, Brazil, 1980.
95. T. J. Slaga, A. J. P. Klein-Szanto, L. L. Triplett, and L. P. Yotti, Skin tumor-promoting activity of benzoyl peroxide, a widely used free radical-generating compound, *Science 213*:1023, 1981.

96. M. K. Logani, C. P. Sambuco, P. D. Forbes, and R. E. Davies, Skin-tumour promoting activity of methyl ethyl ketone peroxide—A potent lipid-peroxidizing agent, *Fd. Chem. Toxic. 22*(11):879, 1984.
97. M. Pathak and T. B. Fitzpatrick, The role of natural photoprotective agents in human skin, In *Sunlight and Man*, T. B. Fitzpatrick, M. A. Pathak, L. C. Harber, M. Seiji, and A. Kukita, Eds., University of Tokyo Press, Tokyo, 1974, p. 725.
98. J. M. Knox, A. C. Griffin, and R. E. Hakim, Protection from ultraviolet carcinogenesis, *J. Invest. Derm. 34*:51, 1960.
99. L. H. Kligman, F. J. Aikin, and A. M. Kligman, Sunscreens prevent ultraviolet photocarcinogenesis, *J. Amer. Acad. Derm. 3*:30, 1980.
100. D. W. Owens and J. M. Knox, Influence of heat, wind and humidity on ultraviolet radiation injury, *Natl. Cancer Inst. Monogr. 50*:161, 1978.
101. C. F. Chignell, R. H. Sik, B. C. Gladen, and D. B. Feldman, The effect of different types of fluorescent lighting on reproduction and tumor development in the C3H mouse, *Photochem. Photobiol. 34*:617, 1981.
102. G. E. Pfahler, J. V. Klauder, and J. L. Martin, Experimental studies on the combined effects of roentgen rays and ultraviolet rays, *Amer. J. Radiol. 16*:150, 1926.
103. E. A. Pohle and C. S. Wright, Studies on the roentgen erythema of the human skin. III. Macroscopic and skin capillary changes after combined exposure to roentgen rays and ultra-violet rays, *Radiology 14*:351, 1930.
104. R. S. Stern, Risk of cutaneous carcinoma in patients treated with oral methoxalen photochemotherapy for psoriasis, *New Engl. J. Med. 360*:809, 1979.

5

SUN PROTECTION FACTORS: REGULATORY ISSUES AND METHODS OF DETERMINATION

JOSEPH J. P. MORTON, Ph.D.
Gillette Medical Evaluation Laboratories
Rockville, Maryland

EMALEE G. MURPHY, Esq.
Cosmetic, Toiletry and Fragrance Association, Inc.
Washington, D.C.

Abstract

Clinical procedures to quantify sunscreen efficacy have been formally developed by government or private standards organizations in the United States, Germany, and Australia. The background of these standards is reviewed. Their chief provisions, including major areas of difference, are described. The radiation source employed is one of the major differences in these methods, and a review of studies evaluating the attributes of natural and artificial radiation sources is provided. Nonclinical test methods are discussed: *in vivo* methods include erythemal and DNA synthesis evaluations; *in vitro* methods involve spectrophotometric analyses and include the dilute-solution, thin-film, and mouse epidermis methods.

Introduction

Methods to determine the protective power of sunscreen products have been discussed in the scientific community since early in this century. Recently, in response to the public's increased awareness of sunscreen benefits, clinical procedures to quantify sunscreen effectiveness have been formally developed by government or private standards organizations in the United States, Germany, and Australia. In general, the purpose of all three methods is to provide the consumer with a useful comparison of sunscreen product efficacy based on standardized test procedures and product labeling. Although some attempts have been made to standardize methodology and thereby to provide a worldwide index of sunscreen product protection, differences among the U.S., Australian, and German test procedures remain.

This paper will discuss these differences and the factors that affect sunscreen product evaluation, including a review of new *in vivo* and *in vitro* methods that have not yet been incorporated into formal standards.

Background

The U.S. Food and Drug Administration's (FDA) Advance Notice of Proposed Rulemaking (ANPR) for over-the-counter (OTC) sunscreen drug products for human use was published in the August 25, 1978 *Federal Register*.[1] The proposal contained the unaltered conclusions of the FDA Advisory Review Panel on Topical Analgesic, Anti-Rheumatic, Otic, Burn and Sunburn Treatment and Prevention OTC Drugs.* This committee was first convened in 1973 as part of the Agency's review of OTC drug safety, effectiveness, and labeling.† Thus, the proposal is not a final regulation, since the FDA must publish its views in a Proposed Rule and must consider further public comment before issuing any Final Rule. Nevertheless, the FDA has stated that marketing a product with a formulation or labeling that is not in accord with a ANPR or Proposed Rule may result in regulatory action against the product, marketer, or both,[2] and most companies marketing sunscreens in the United States currently use the proposal as a guide for sunscreen composition, testing, and labeling.

The Panel considered 21 of the 27 sunscreen agents reviewed as safe and effective and not misbranded, if used at recommended concentration levels and provided they exhibit a sun protection factor (SPF) value of at least 2. The SPF is defined as the

*Panel members included Thomas G. Kantor, M.D., Chairman, John Adriani, M.D., Col. William A. Akers, M.D., Maxine Bennet, M.D., Minerva S. Buerk, M.D., Walter L. Dickison, Ph.D., and Jerry H. Shuck, M.D. Nonvoting representatives of consumer organizations, the Proprietary Association, The Cosmetic, Toiletry and Fragrance Association, and the Food and Drug Administration served as panel liaisons and administrative staff.

†The OTC drug review, authorized in 1962 by the Harris-Kefauver Amendment to the Federal, Food, Drug and Cosmetic Act (Public Law 87-781, 87th Congress, S.1552, October 10, 1962) was first announced by FDA in the January 5, 1972 *Federal Register* (37 *Fed. Reg.* 85).

UV energy required to produce a minimal erythema dose (MED) on protected skin divided by the UV energy required to produce MED on unprotected skin. (An individual MED is the exposure time necessary to produce a minimally perceptible erythema at 16–24 hr after radiation.) Furthermore, the Panel proposed a system for labeling sunscreen products using SPF values to indicate that the user could stay "*x*" times longer in the sun without burning than would be possible if the sunscreen were not used.

To help consumers choose the appropriate sunscreen products, the Panel created descriptive categories of common sunburn and tanning histories by which consumers could classify their own skin types. The Panel also recommended five product category designations based on SPF values obtained by the proposed test methodology (see Table 1).

The German methodology (formulated by the Deutsches Institut fuer Normung e.v.: DIN) for ranking sunscreens does not mention product categories, but in Australia, four sunscreen product categories are specified: minimum, SPF 2–4; moderate, SPF 4–8; high, SPF 8–15; and maximum, SPF 15 and above.

Table 1. Product Category Designations Based on SPF Values

Skin Type	Description	SPF	Category Descripton
I	Always burns easily, never tans	15 or greater	*Ultra*: most protection, no suntanning
		8–14	*Maximal*: maximal protection, little or no suntanning
II	Always burns easily, tans minimally	6–7	*Extra*: Extra protection from sunburn and permits limited tanning
III	Burns moderately, tans gradually (light brown)	4–5	*Moderate*: Moderate protection from sunburning, permits some suntanning
IV	Burns minimally, always tans well (moderate brown)	2–3	*Minimal*: protection from sunburning, permits suntanning
V	Rarely burns, tans profusely (dark brown)	2	*Minimal*
VI	Never burns, deeply pigmented	None	None indicated

In adopting the sun protection factor rating system for sunscreen effectiveness, the Panel recognized the need for a standardized method for quantifying the protective power of sunscreen products. The use of an index to indicate sunscreen efficacy was not unknown to Panel members, since protective indices had been used previously in Europe and discussed in U.S. scientific journals.[3, 4] However, the lack of standardized procedures to determine SPF values often made accurate comparison of commercial sunscreen products difficult.[4-6] The Panel therefore proposed testing procedures to reduce avoidable variation in measured SPF values.

In Germany, discussion of a standardized method to determine sunscreen product protection began in 1980 under the auspices of the private German standards organization, DIN. The responsible DIN Committee, Normenausschuss Lichttechnik (Technical Light Committee), issued its draft propcsal for comment in March, 1984, "Evaluation of Sunscreen Products; Experimental Dermatological Evaluation of the Protection Against Erythema of External Sunscreen Products for the Human Skin" (DIN 67501). This draft incorporated many of the techniques developed in earlier German studies and differed in many respects from the FDA's proposed methodology. As part of the comment process, industry association representatives from the United States, France, Great Britain, and Germany met during the summer of 1984 in an attempt to avoid significant differences that could result in disparate SPF values for tested sunscreen products.* Formal comments to the DIN were subsequently submitted by the American, German, and European Economic Community (EEC) cosmetic trade associations.† The final DIN standard was issued in September, 1984 with a 1-year delay in the effective date. The standard incorporates several changes suggested by the industry, although key differences between DIN and FDA test procedures remain.

In Australia, similar concerns about the need for a simple, quantitative method for determining the protective power of sunscreen products resulted in a request from the Commonwealth Department of Health to the Standards Association of Australia (SAA), a private organization, to develop appropriate methodology. The SAA Committee on Sunscreen Agents (CS/42) subsequently issued the Australian Standard for Sunscreen Products—Evaluation and Labelling (AS 2604-83) as a guide to sunscreen labeling for companies marketing these products in Australia. A revised draft incorporating additional standards for water-resistant products, other products such as make-up containing sunscreen (secondary sunscreen products), and products designed to protect against both UVA and UVB radiation (broad-spectrum protection products)

*Representatives of The Cosmetic, Toiletry and Fragrance Association (U.S.), The Cosmetic, Toiletry and Perfumery Association (U.K.), the Federation Française de l'Industrie des Produits de Parfumerie, de Beauté, et de Toilette (France) and the Industrieverband Koerperpflege und Waschmittel (Germany).

†Letter from CTFA President E. Edward Kavanaugh, to Normenausschuss Lichttechnik (DIN), June 25, 1984; Letter from Dr. A. Schlitt and Dr. R. Schnakig (IKW) to Normenausschuss Lichttechnik (DIN), June 22, 1984; COLIPA (Comité de Liaison des Associations Européenes de l'Industrie, de la Perfumerie, des Produits Cosmetiques et de Toilette) Comments to Normenausschuss Lichttechnik (DIN), doc. 84/218, June 1984.

was issued for comment in November, 1984. Formal comments on the revised draft were submitted by the U.S. Cosmetic, Toiletry and Fragrance Association and COLIPA, the federation of cosmetic trade associations within the European Economic Community.* Although the Australian methodology under discussion is similar to the FDA's proposal, there is not complete standardization with either the FDA or DIN methodology.

The legal status of the U.S., German, and Australian standards also differs and the practical impact for industry remains somewhat unclear. The FDA's proposal must be issued again for comment in a Proposed Rule before a final regulation can be published. Until then, the proposal will continue to operate as a guide for the industry's determination of sunscreen product efficacy. The FDA may also choose to publish test methodology as a guideline that companies may use, rather than as a formal regulation, as it has with procedures for testing antiperspirant efficacy.[7] The German and Australian standards, although issued by private standards organizations, may be used by the respective governments to assess the validity of sunscreen product labeling. In fact, the SAA has developed a certification procedure and mark to identify products tested and labeled in conformance with the SAA standard.

The publication of standard sunscreen testing methods in the United States, Germany, and Australia has prompted discussion of sunscreen efficacy in other countries, including Japan and Canada. While not formally adopting any one method, the Canadian Health Protection Branch issued a technical information bulletin designed to clarify the difference between sun protection products marketed as drugs and tanning products marketed as cosmetics.[8] The Canadian approach to "sun product" labeling differs in at least three areas from the approach taken by the FDA. First, claims for the prevention of burns or other protection from the effects of the sun, including SPF labeling, should not be made for cosmetic products. To support a drug claim for a sunscreen product, the product must have an SPF value of greater than 4. The FDA's proposal permits drug sunscreen labeling for products with SPF values of 2 or greater.

Second, although cosmetic products may not use SPF values in their labeling or advertising, the Health Protection Branch will permit claims that a cosmetic product will allow "*x*" times the normal exposure time to the sun, provided no references to sunburn or sunburn prevention are made. The FDA, in contrast, generally considers any reference to SPF values as labeling appropriate for drug products only.

Third, the Health Protection Branch considers sunscreen drug claims such as "extra," "maximal," and "ultra" to be inappropriate and all references to skin cancer are prohibited, although a sunscreen drug product may claim that liberal and regular use "may help to reduce the chance of premature aging of the skin." "Ultra," "maximal," and "extra" designations for sunscreen products are, however, included in

*Letter from E. G. Murphy (CTFA) to James Weeks, Cosmetic, Toiletry and Fragrance Association of Australia, February 13, 1985; Letter from Jack Dupuis (COLIPA) to James Weeks, Cosmetic, Toiletry and Fragrance Association of Australia, January 31, 1985.

the FDA Panel proposal, as well as the claim that "overexposure to the sun may lead to premature aging of the skin and skin cancer. The liberal and regular use over the years of this product may help reduce the chance of premature aging of the skin and skin cancer."

In Japan, a special committee of the Japan Cosmetic Industry Association is currently discussing appropriate test methodology for sunscreen products, including both guinea pig and human clinical methods.* Since a final proposal has not yet issued from these discussions, Japan's methodology is not discussed in detail here.

Correspondence among industry groups in the United States, European Economic Community, Japan, and Australia confirms a general desire to establish compatible, if not identical, test methods. However, to date, such has not been achieved.

Comparison of FDA, DIN, and SAA *In Vivo* Human Test Methods

The chief provisions of the FDA proposal and DIN and SAA standards are described in the following sections.

Standard Reference Formulation

FDA

A standard reference formulation of 8% homomenthyl salicylate (homosalate) with a known SPF value is used to verify test procedures. The original reference standard published in the August 28, 1978, *Federal Register* has, in practice, been replaced by a revised formulation (SPF 4.11 by FDA method) validated by an interlaboratory test in four laboratories, the formulation for which is:

	Ingredients	Percentage
Part A	Lanolin	5.00
	Homomenthyl salicylate	8.00
	White petrolatum	2.50
	Stearic acid	4.00
	Propylparaben	0.05
Part B	Methylparaben	0.10
	EDTA disodium	0.05
	Propylene glycol	5.00
	Triethanolamine	1.00
	Purified water	74.30

*Comments of the Japan Cosmetic Industry Association on Sun Protection Factor, March 1, 1985, to CTFA.

DIN

A reference standard sunscreen preparation of 2.70% *p*-methoxy-cinnamic-acid-2-ethylhexylester (SPF 3.7 ± 0.3 by DIN method) is used to verify test procedures. The formulation is:

	Ingredients	Percentage
Part A	Tegin P (Th. Goldschmidt)	1.00
	Liquid paraffin (80 mPas at 20°C)	5.00
	Triple-pressed stearic acid (stearic acid app. 45%; palmitic acid app. 50%)	1.50
	P-methoxy-cinnamic acid-2-ethylhexylester	2.70
	Lanette O (Henkel)	0.40
	Propyl paraben	0.10
Part B	Pure triethanolamine	0.80
	Glycerine (85%)	4.00
	Methyl paraben	0.10
	Carbopol 934 P (Goodrich)	0.10
	Demineralized water	84.30

SAA

The reference standard is the revised formulation used by the FDA, which companies may use to validate test procedures.

Test Panel

FDA

Each test panel should include at least 20 persons (male and female) with skin types I, II, or III. Skin types described by the FDA Panel are outlined in Table 1. Individuals with scars or blemishes in the test site area or with medication-induced or naturally abnormal responses to sunlight are eliminated. Informed consent must be obtained from all panelists.

DIN

Test panels must include at least 20 persons whose UV sensitivity is representative of the majority of sunscreen users. Tanned individuals are excluded as test subjects. (The responsibility for excluding other inappropriate test subjects is left to the researcher, usually a dermatologist.)

SAA

Test panels should include a minimum of 10 persons with skin types I, II, or III. Volunteers with a history of abnormal response to medication or UV radiation, or allergies to topically applied cosmetics should be excluded. The Australian Health and Medical Research Council recommends obtaining informed consent from each test subject.

Test Area

FDA

The test area should be on the back between the beltline and shoulder blade and lateral to the midline. The minimum test area is 50 cm^2. Individual test sites should be at least 1 cm^2.

DIN

The test area should not exceed 30 cm $\times$ 30 cm (900 cm^2), located on both sides of the spine, and subdivided into test sites of at least 0.4 cm^2 separated by 1-cm wide UV-opaque plaster strips. Each test strip must be adjacent to a control strip.

SAA

The minimum test area, which is not limited to the back, is 30 cm^2. Individual test sites should be not less than 1 cm^2.

Application of Test Material

FDA

Test materials are applied at 2 mg or 2 $\mu l/cm^2$.

DIN

Test materials are applied at 1.5 $\pm$ 10% mg/cm^2.

SAA

Test materials are applied at 2 mg/cm^2.

Waiting Period

FDA

At least 15 min must elapse before subjects are exposed to natural sunlight or artificial light source.

DIN

Radiation exposure should begin 20 min following application of test materials.

SAA

At least 15 min should elapse before exposure. Where specified, the product should be allowed to dry in accordance with the manufacturer's instructions.

Artificial Light Source

FDA

The preferred light source is the xenon arc solar simulator,[9] defined as having a continuous emission spectrum in the UVB (290–320-nm) range, with less than 1% of its total energy contributed by nonsolar wavelengths (shorter than 290 nm) and not more than 5% of its erythemogenically effective energy contributed by nonsolar wavelengths. Spectral shaping is achieved using UVB and infrared filters. The unit must be monitored periodically to assure delivery of the appropriate radiation spectrum. Other light sources may be used as long as results can be verified by corroborative testing using sunlight.

DIN

The preferred artificial light sources are the Osram Ultravitalux or Philips MLU 300-W-E lamps. Other light sources may be used as long as they yield comparable SPF values (no more than ± 20%) when tested using at least three sunscreens of high, medium, and low sun protective effectiveness.

SAA

The xenon arc solar simulator is the preferred artificial light source, defined generally in accordance with the FDA description but including continuation of the spectrum through the UVA region, up to 400 nm.

Sunlight As Radiation Source

FDA

Proposed methodology for sunlight testing specifies that subjects lie prone in direct sunlight for a predetermined period of time. Sun exposure for each individual subject must be completed during one continuous exposure period, although the day of sun exposure may not be the same for all subjects. Sun exposure of all subjects for any one test must be at the same geographical location and must be completed within 2 weeks. Test site exposure is controlled using a grid or template of opaque materials; the remainder of the back is covered with similar opaque material or toweling.

During each exposure, the sun intensity should be measured continuously and the duration of sun exposure should be documented in J/m^2 or in Robertson-Berger meter counts. Temperature and humidity should be measured at the beginning, end, and at the maximal sun intensity for the exposure period. Primary means to measure variations in cloud cover during exposure are a continuous radiometer or Robertson-Berger meter record, although descriptions of wind and cloud conditions may be made. Test subsite areas are exposed at preestablished times as determined by the meter reading to obtain graded exposures and an identical sequence of exposures should be administered to all test sites. The SPF value of the tested sunscreen product using the Robertson-Berger meter is calculated as the exposure count interval (MED-protected skin) divided by the exposure count interval (MED-unprotected skin).

DIN

DIN 67501 does not include guidelines for sunlight testing.

SAA

Natural sunlight is considered too variable and unpredictable to be used routinely for assessing large numbers of sunscreen products.

Exposure Times and Readings

FDA

Radiation exposure progresses geometrically so that the length of time for each succeeding exposure is 25% longer than the previous one. Results are read at between 16 and 24 hr following irradiation.

DIN

Exposure progresses geometrically so that the length of time for each succeeding exposure is 40% longer than the previous one. Results are read at between 20 and 28 hr following irradiation.

SAA

Exposure time increases geometrically so that the length of time for each succeeding exposure is no more than 26% longer than the previous one. Results are read at between 16 and 24 hr after irradiation.

Data Evaluation

FDA

The sunscreen product SPF is the arithmetic mean of the individual SPF test values. The standard error must not exceed $\pm$ 5% of the mean.

DIN

The individual data are transformed to $\log_{10}$ values and the sunscreen product SPF is the geometric mean of the individual SPF test values.

SAA

Data are calculated using the same guidelines as the FDA. The standard error of the mean must not be greater than 10% of the mean protection factor.

Light Source

The light source is one of the variables in the several methods used for SPF determinations. Sunlight, under conditions of normal or controlled use, would be the obvious light source. Several recognized factors influence the determinations of SPF values under normal use conditions, including temperature, high or low humidity, direct and scattered radiation, snow, sand, wind velocity, and sweating.[11-13] Controlled tests that use sunlight, although less variable, still present difficulties, including considerations of the season, latitude, altitude, clouds, smog, and time of day. Since it is difficult to control these numerous variables, tests are not generally conducted using sunlight. These variables are recognized in the SAA method, which notes that ". . . although sunlight may at times be convenient for product testing, it is too variable and unpredictable to be used routinely for assessing large numbers of sunscreen products." For these reasons, *in vivo* human tests more commonly employ an artificial light source.

The FDA and SAA methods recommend use of the xenon arc solar simulator. The FDA method requires a radiation source with a continuous emission spectrum in the UVB range of 290–320 nm. The SAA method requires a continuation of the spectrum through the UVA region, up to 400 nm. This upper range is not highly erythemogenic and the extension of this range is not expected to alter SPF determinations substantially when tested using the *in vivo* human protocol. The SAA method evaluates UVA efficacy (broad-spectrum products) using an *in vitro* test method. Figure 1 shows the spectrum of natural sunlight and a xenon arc radiation source. The energy emitted from the xenon arc source is considered to be constant, and the MED is determined by varying the exposure time. The required range of 290–320 nm (FDA method) is attained using appropriate filters. Extending the lower limits of this highly erythemogenic region can result in different SPF values. In an *in vivo* human study[14] a WG-305 filter, 2-mm thick, was compared to a WG-320 filter, 1-mm thick, using the same xenon arc light source. The WG-305 filter extends the range to 280 nm. The authors reported a significantly higher SPF value with the WG-305 filter when 4% PABA was evaluated (SPF of 5.72 vs. 4.56). However, a similar increase in the SPF value was not noted for an 8% homosalate formulation; this difference my be due in part to differences in the spectral absorption properties of the two sunscreens.

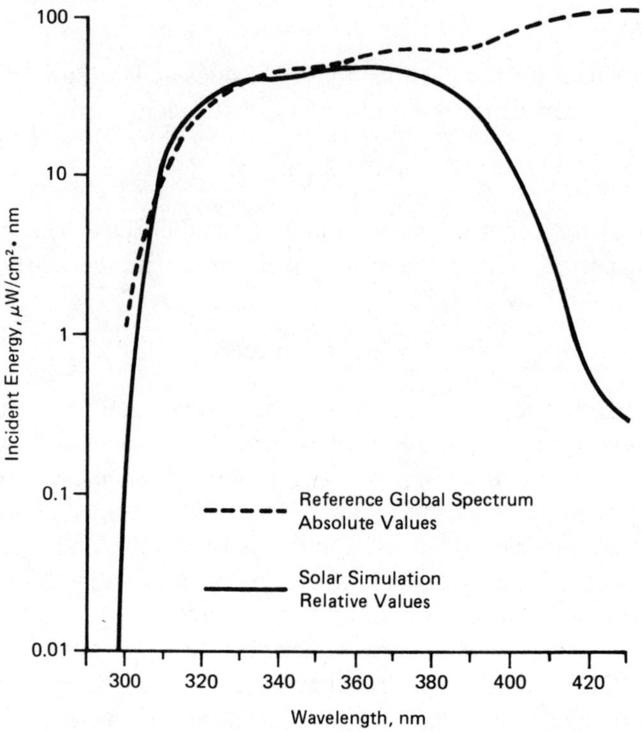

Figure 1. Solar simulator. Reprinted with permission from reference 10.

The Osram Ultravitalux lamp (an intermediate-pressure filtered mercury vapor lamp) is one of the preferred light sources in the DIN method. Maximum energy is emitted at wavelengths of 297, 303, 313, and 365 nm. Cripps[15] has studied somewhat similar wavelengths using monochromatic radiation of 295, 300, 305, and 313 nm. Four commercially available sunscreen products were tested in humans at each of the wavelengths. The resulting SPFs were essentially comparable at each of these wavelengths, even though a 100-fold or greater increase in dosage (μW sec/cm^2) was necessary to produce erythema at the higher wavelengths.

Recently, Sayre and Agin compared SPF values obtained using a xenon arc solar simulator and the Ultravitalux lamp.[16] Both *in vivo* and *in vitro* protocols were used. The *in vivo* test in humans employed the solar simulator in accordance with the FDA method. The *in vitro* method employed both the solar simulator and the Ultravitalux lamp. The solar simulator in the *in vitro* studies was fitted with either a 1-mm WG-320 or a 2-mm WG-305 filter. The *in vitro* method determined the SPF value by a forward-scattering technique conducted over the range of 250–400 nm. This *in vitro* technique is further discussed below. Twelve products with commercially labeled SPF categories ranging from 2 to 15 were evaluated. The predicted protection factors determined from *in vitro* testing and the protection factors determined from *in vivo* human testing are listed in Table 2. With but a few exceptions, all of the resulting SPF values are quite similar. The authors concluded that any differences noted in SPF determinations using the FDA method and the DIN method would not be expected to be a result of the spectral differences between the solar simulator and the Ultravitalux lamps. The authors attribute this similarity in part to the use of multiple sunscreen agents in products. Such "broad-spectrum" products increase the range

Table 2. Comparison of SPFs of Sunscreen Formulations Using various Light Sources

Active Ingredients	Predicted WG320 Xenon Arc PF	Predicted WG305 Xenon Arc PF	Predicted Osram PF	Human PF
A. 8% Padimate 0[a] 3% Oxybenzone	16.0 ± 3.0	17.9 ± 4.2	17.1 ± 4.1	15.3 ± 2.2
B. 5% Padimate 0	6.3 ± 0.9	6.0 ± 0.8	5.9 ± 0.8	6.5 ± 0.6
C. 6% Padimate 0[a] 2% Oxybenzone	8.3 ± 2.1	9.3 ± 3.0	8.9 ± 2.8	8.9 ± 1.1
D. 7.5% 2-ethylhexylpara-methoxycinnamate 3% Oxybenzone	16.4 ± 4.2	21.1 ± 6.3	20.9 ± 6.3	15.9 ± 2.4
E. 8% Padimate 0[a] 3% Oxybenzone	16.9 ± 1.0	20.2 ± 1.8	19.6 ± 1.7	18.1 ± 1.9
F. 6% Padimate 0[a] 2% Oxybenzone	9.2 ± 0.5	10.1 ± 0.2	9.8 ± 0.3	8.9 ± 1.2
G. 4% Homosalate	1.5 ± 0.3	1.4 ± 0.3	1.4 ± 0.3	2.1 ± 0.3
H. 8% Homosalate	3.8 ± 0.2	3.5 ± 0.2	3.4 ± 0.2	4.0 ± 0.6
I. 3.8% Padimate 0	3.9 ± 1.0	4.9 ± 1.6	4.7 ± 1.5	5.5 ± 1.6
J. 5.3% Padimate 0 1.75% Oxybenzone	8.6 ± 0.2	10.8 ± 0.2	10.4 ± 0.2	9.8 ± 1.8
K. 2.7% 2-ethylhexylpara-methoxycinnamate	3.5 ± 0.3	3.7 ± 0.3	3.7 ± 0.3	3.8 ± 0.5
L. 5% Paraaminobenzoic acid	5.7 ± 1.3	12.1 ± 3.1	11.4 ± 2.9	12.01 ± 3.11

Arithmetic mean values ± standard deviations based upon the use of small population statistics.
[a]C and F contained similar active ingredients, but were each unique product formulas, as are A and E.
Reprinted with permission from ref. 16.

of spectral protection and cause the efficacy to be relatively insensitive to the spectral distribution of the radiation source.

Cole and colleagues have also addressed the relationship between the spectral absorption properties of sunscreen agents, the spectral distribution of the source of radiation, and protection factors.[17] These investigators determined the predictive protection factors of several sunscreens using a mercury lamp and a xenon arc solar simulator. The mercury lamp used emitted maximum energy at wavelengths of 297, 303, 313, and 365 nm, which are the same as noted for the Ultravitalux lamp. The predictive protection factors are calculations based on the spectral irradiance from the two radiation sources, an erythema action spectrum, and spectral transmission properties of various sunscreens in solution. Cole concluded that products with a maximum absorbance in the 300–320-nm band will have a higher protection factor when tested with a xenon lamp. Conversely, products with maximum absorbance between 295 and 305 nm will have a higher protection factor when tested with the mercury lamp. For "broad-spectrum" products exhibiting a relatively flat absorbance spectrum between 290 and 320 nm, the protection factors will be aproximately the same for the two radiation sources.

The Cosmetic, Toiletry and Fragrance Association has recently begun an interested party research program designed to qualify the xenon arc lamp systems used by many U.S. companies for use with DIN methodology. The program involves comparing SPF values obtained using both lamps under the same human test conditions. Either the FDA or DIN preferred lamps might, in theory, be acceptable should testing reveal no significant lamp-derived differences in SPF values.

Animal Methods

In vivo tests in animals are not currently in widespread use; however, the hairless mouse has been used.[18] In this model, 1 cm^2 of skin is irradiated with a xenon arc solar simulator. Test groups of animals are treated with the sunscreen product followed by irradiation of 2.5, 5, 7.5, or 10 MED. Control animals are similarly irradiated without prior application of the sunscreen product. The study noted that the skin reactions were reproducible. However, the response to UV radiation appeared to be more intense and may have involved damage to deeper layers of the skin than would have been noted in humans. This enhanced response was thought to be due to a thinner horny layer, a thinner epidermis, and the lack of epidermal pigment in the hairless mouse than in humans. The authors note that the hairless mouse may be a promising model for comparing the effectiveness of sunscreen products on skin *in vivo*. Other investigators have noted that the evaluation and accurate grading of UV-induced erythema are difficult in hairless mice due in part to dermal edema.[19]

More recently, several researchers have reported use of an *in vivo* mouse model based on the earlier findings of Epstein and colleagues.[20] These investigators noted that ultraviolet rays ($<$ 320 nm) inhibit DNA, RNA, and protein synthesis in hairless mouse skin *in vivo* within 1–2 hr after exposure. The inhibition persists for at least 16 hr. Recovery is noted by 24 hr, with acceleration of these functions reaching a peak at 48–72 hr. Lowe and colleagues noted similar findings.[19] Epidermal DNA synthesis decreased at 6 and 24 hr and increased to a maximum 48 hours after irradiation. These changes were preceded by increases in ornithine decarboxylase activity at 2 hr, an 80-fold increase at 24 hr, and a decrease at 48 hr.

Several investigators have used these observed changes in epidermal DNA synthesis for the *in vivo* evaluation of sunscreen efficacy. One method involves irradiating (240–325 nm) groups of hairless mice with and without prior sunscreen application.[21] At either 6 or 48 hr after application, the animals are sacrificed. One hour before sacrifice, the animals are injected with 25 Ci tritiated thymidine. The epidermal specimens are excised, homogenated, extracted using hydroxyapatite column, and expressed as counts per minute per μg of DNA. Six hours after irradiation, high sunscreen protection is denoted by normal or only slightly decreased levels of DNA synthesis and low sunscreen protection is denoted by suppressed DNA synthesis. Forty-eight hours after irradiation, high sunscreen protection is denoted by normal or only slightly increased levels of DNA synthesis and low sunscreen protection is denoted by an increase in DNA synthesis.

Walters and Dequoy used a similar method but evaluated DNA synthesis activity using a direct disc technique instead of column extraction.[22, 23] They evaluated DNA synthesis activity 2 hr after irradiation by determining the amount of UV (275–380 nm; peak at 313 nm) that inhibits DNA synthesis by 50%. Figure 2 shows the effect

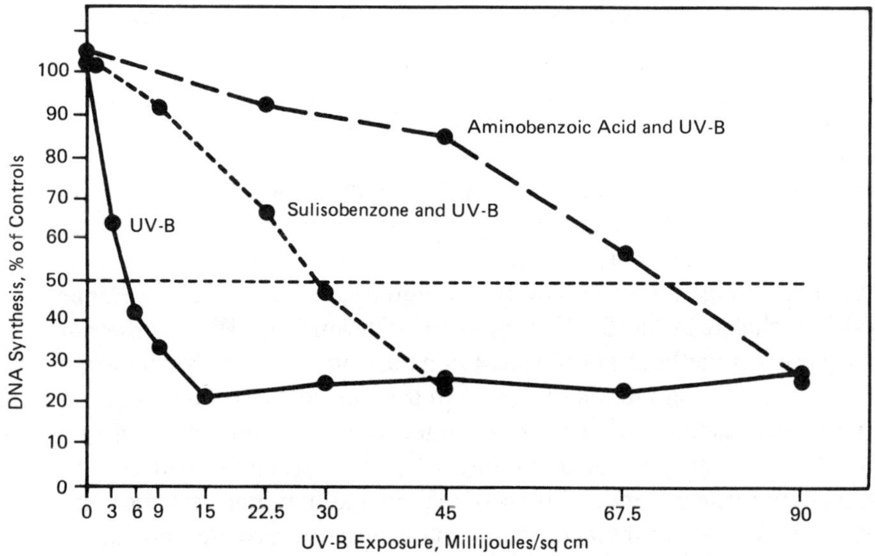

Figure 2. Effect of various doses of UVB radiation on epidermal DNA synthesis with and without sunscreens. Reprinted with permission from reference 22.

Table 3. Comparison of Sunscreen Protection by Different Methods

Sunscreen Product	SPF Label Claim	SPF Human *In Vivo*	PF Mouse *In Vivo*
Coppertone 4	4	4.0[a]	4.4
Sundown Extra Protection	6	5.9[b]	8.4
Supershade 15	15	17.4[a]	21.0
Eclipse 15	15	15.2[a]	22.2
Blockout 15	15	15.9[a]	22.4
Bain de Soleil 15	15	23.9[a]	27.6

[a]Actual SPF claimed by manufacturer.
[b]Datum from ref. 29.
Reprinted with permission from ref. 24.

of various doses of UVB radiation on epidermal DNA synthesis with and without sunscreens (aminobenzoic acid and sulisobenzone). The estimated point of 50% inhibition of DNA synthesis is 5 mJ/cm^2 for untreated skin. Increasing UVB exposure beyond 15 mJ/cm^2 does not further decrease DNA synthesis. In another study seven sunscreens were evaluated by this method and their protection factors were compared to human test results.[24] As noted in Table 3, the mouse *in vivo* PF values are somewhat higher than the values obtained in humans. The authors note that these differences could be due to differences in light sources, the dosages of sunscreen applied, and the arbitrary use of 50% inhibition of DNA synthesis as the end point for calculation of the SPF. The authors' assessment is that the "model can only serve as a rough guide to relative sunscreen UVB protection before final sunscreen testing in humans with a solar simulator or the sun as the radiation source."

In Vitro Test Methods

The two most widely used *in vitro* methods employ spectrophotometric analyses. One method involves measuring the transmission characteristics of a dilute solution of sunscreen material in a suitable solvent.[25] The second method measures UV transmission through a thin film of the undiluted sunscreen material.[26] The thin-film method is included in the SAA method for determining UVA sunscreen efficacy. Both of these test methods are physical in nature and do not consider any of the factors involved in the interaction of skin with the sunscreen product. This may explain in part why the results from these *in vitro* techniques, although precise, do not correlate well with human test data obtained using an artificial radiation source or SPF determinations made under conditions of use. In addition, both methods have inherent limitations.[27] Because of their solubility characteristics, products containing physical sunscreens such as titanium dioxide and zinc oxide are not suitable for evaluation by the dilute solution method. The desired thickness and uniformity of the film in the thin film method are difficult to attain consistently.

Table 4. Comparison of the Sun Protection Factors Predicted by the Solution Method to the SPFs Obtained in Human Testing

Product	SPF From Human Testing	SPF Predicted in Isopropanol	Difference in Orders of Magnitude
A	1.74×10^1	3.9×10^{18}	10^{17}
C	1.3×10^1	1.4×10^9	10^8
G	1.2×10^1	2.8×10^2	10^1
B	8.2×10^0	8.0×10^7	10^7
E	8.1×10^0	7.9×10^8	10^8
D	6.7×10^0	6.5×10^6	10^6
F	4.6×10^0	6.6×10^5	10^5

Reprinted with permission from ref. 27.

Sayre and colleagues have developed an additional *in vitro* method that compensates in part for the deficiencies of the thin film method.[27] It modifies the basic thin-film method by using excised hairless mouse epidermis positioned on a quartz plate. Forward-scattering scans are run from 250 to 450 nm with a recording spectrophotometer fitted with a diffuse reflectance sphere. The forward-scattering scan is repeated following the application of sunscreen product to the epidermis at a density of 2 μL/cm^2. The predicted SPF is a function of the determined transmission spectrum, the spectrum of the light source, and the erythemogenic efficiency spectrum. These investigators tested seven commercially available sunscreen products using the dilute-solution method, a human method similar to the FDA method, and the mouse epidermis method. The dosage in each study was 2 μL/cm^2 or the equivalent. Table 4 compares the SPF values determined from human testing and the dilution solution method. As noted, the dilution solution method gave SPF values much greater than those determined from human testing. Table 5 compares the same human test results with those obtained from the mouse epidermis method. The authors note that the two sets of SPF determinations fall within comparable statistical ranges.

Table 5. Comparison of Sun Protection Factors Obtained by Human Testing and Hairless Mouse Epidermis

	SPF	
Product	Human Testing	Hairless Mouse Skin
A	17.41 ± 3.17	22.8 ± 10.1
C	13.03 ± 4.09	12.4 ± 9.3
G	12.01 ± 3.11	6.8 ± 1.5
B	8.20 ± 2.69	10.2 ± 0.02
E	8.13 ± 2.42	9.5 ± 4.9
D	6.70 ± 1.65	6.8 ± 1.2
F	4.60 ± 1.02	3.4 ± 0.5

Reprinted with permission from ref. 27.

A second similar study was conducted on nine commercially available products marketed in Australia.[28] The thin-film method was used in addition to the dilute solution method, the mouse epidermis method, and the human test method. Similar results were observed in all. It was concluded that the mouse epidermis method was the *in vitro* method of choice since it gave SPF values comparable to those from human testing. The authors note that the results "appear to indicate that sunscreen preparations undergo physical and possible chemical interactions with the components of the epidermis that may lead to an entirely different absorption spectrum than is depicted by spectrophotometric methods of evaluation alone." The authors regard the mouse epidermis method as superior to any previously reported *in vitro* technique for testing sunscreen effectiveness. They further note that "although *in vivo* procedures are to be preferred as a final evaluation, the results obtained in this study would indicate that the hairless mouse epidermis method could serve as a reliable replacement."

Conclusions

Discussion of appropriate methods to determine sunscreen efficacy continues. The Food and Drug Administration has recently indicated its interest in reviewing the methodology proposed in 1978 and has solicited industry comment on current industry practices. No doubt, the views of industry and academia, as well as consumer needs, will be reflected in any final FDA treatment of SPF determination testing. Similarly, discussion in Australia continues on evaluation of "broad-spectrum" sunscreens designed to protect against UVA, as well as UVB, radiation. Determination of SPF is an evolving science and recent or continuing studies may help point the way to clearer understanding of the meaning and usefulness of SPF values and SPF product labeling. Animal *in vivo* and *in vitro* testing appears to provide a useful guide to product research and development, but human testing at this time remains the accepted method for determining the SPF values of sunscreen products.

References

1. 43 *Federal Register* 8206.
2. 21 Code of Federal Regulations 330.13(b)(2).
3. F. Greiter, P. Bilek, and S. Doskoczil, History of sunscreens and the rationale for their use, In *Principles of Cosmetics for the Dermatologist*, P. Frost and S. Horowitz, Eds., C. V. Mosby, St. Louis, 1982.
4. I. Willis and A. Kligman, Evaluation of sunscreens by human assay, *J. Soc. Cosmet. Chem.* *20*:639–651, 1969.
5. R. Schulz, Some tests and remarks on the problem of commercial sunscreens, *Parfueme Kosmetik*, 1956.
6. U. Hoppe, H. J. Kopplow, and A. Wiskemann, Statistical evaluation of the light protection-factor, *Drug Res.* *25*(5):817, 1979.

7. Antiperspirant Drug Products for Over-the-Counter Human Use; Tentative Final Monograph, 47 *Fed. Reg.* 36492-36504, August 20, 1982.
8. Technical Information Bulletin of the Drugs Directorate, Health Protection Branch, No. 7, November, 1981.
9. D. Berger, Specification and design of solar ultraviolet simulators, *J. Invest. Derm.* *53*(3):192, 1969.
10. Standards Association of Australia, Sunscreen Products, Evaluation and Classification, AS 2604-83, North Sydney, N.S.W., Australia.
11. M. A. Pathak, Sunscreens: Topical and systemic approaches for protection of human skin against harmful effects of solar radiation, *J. Am. Acad. Dermatol.* *7*(3):285, 1982.
12. R. M. Sayre, D. L. Desrochers, E. Marlowe, and F. Urbach, The correlation of indoor solar simulator and natural sunlight, *Arch. Derm.* *114*:1649, 1978.
13. R. M. Sayre, E. Marlowe, P. P. Agin, G. J. LeVee, and E. W. Rosenberg, Performance of six sunscreen formulations on human skin, *Arch. Derm.* *115*:46, 1979.
14. G. J. LeVee, R. M. Sayre, and E. Marlowe, Sunscreen product effectiveness can vary with different simulated solar ultraviolet spectra, *J. Soc. Cosmet. Chem.* *31*:173, 1980.
15. D. J. Cripps and S. Hegedus, Protection factor of sunscreens to monochromatic radiation, *Arch. Derm.* *109*:202, 1974.
16. R. M. Sayre and P. P. Agin, Comparison of human sun protection factors to predicted protection factors using different lamp spectra, *J. Soc. Cosmet. Chem.* *35*:439, 1984.
17. C. Cole, R. E. Davies, and P. D. Forbes, Effect of irradiation source on predicted SPF; Mercury-doped incandescent lamp vs. xenon arc solar simulator, personal communication, 1985.
18. H. Wolska, A. Langner, and F. N. Marzulli, The hairless mouse as an experimental model for evaluating the effectiveness of sunscreen preparations, *J. Soc. Cosmet. Chem.* *25*:639, 1974.
19. N. J. Lowe and J. Breeding, Evaluation of sunscreen protection by measurement of epidermal DNA synthesis, *J. Invest. Derm.* *74*:181, 1980.
20. J. H. Epstein, K. Fukuyama, and K. Fye, Effects of ultraviolet radiation on the mitotic cycle and DNA, RNA and protein synthesis in mammalian epidermis *in vivo, Photochem. Photobiol.* *12*:57, 1970.
21. N. Lowe, A. Verma, and R. K. Boutwell, Ultraviolet light induces epidermal ornithine decarboxylase activity, *J. Invest. Derm.* *71*:417, 1978.
22. J. F. Walter and P. R. DeQuoy, The hairless mouse as a model for evaluating sunscreens; prevention of ultraviolet B inhibition of epidermal DNA synthesis, *Arch. Derm.* *116*:419, 1979.
23. A. S. Otani, R. W. Gange, and J. F. Walter, Epidermal DNA synthesis: A new disc technique for evaluating incorporation of tritiated thymidine, *J. Invest. Derm.* *75*:375, 1980.
24. J. F. Walter, Evaluation of seven sunscreens on hairless mouse skin, *Arch. Derm.* *117*:547, 1981.
25. B. M. Cumpelik, Analytical procedures and evaluation of sunscreens, *J. Soc. Cosmet. Chem.* *23*:333, 1972.
26. D. F. Robertson and G. A. Groves, The selection and use of topical sunscreens, *Med. J. Aust.* *2*:1445, 1972.
27. R. M. Sayre, P. P. Agin, G. J. LeVee, and E. Marlowe, A comparison of *in vivo* and *in vitro* testing of sunscreening formulas, *Photochem. Photobiol.* *29*:559, 1979.
28. G. A. Groves, P. P. Agin, and R. M. Sayre, *In vitro* and *in vivo* methods to define sunscreen protection, *Aust. J. Derm.* *20*:112, 1979.
29. R. M. Sayre, P. P. Agin, D. L. Desrochers, and E. Marlowe, Sunscreen testing methods: *In vitro* predictions of effectiveness, *J. Soc. Cosmet. Chem.* *31*:133, 1980.

6

PHOTOCHEMOTHERAPY

RICHARD S. BERGER, M.D.
GEORGE D. REILLY, M.D.
Division of Dermatology
University of Medicine & Dentistry of New Jersey–Rutgers Medical School
New Brunswick, New Jersey

Abstract

The therapeutic benefit of ultraviolet light in certain diseases, notably psoriasis, with and without the addition of photosensitizing drugs, has been recognized for many years. The combination of a photoactive chemical and light defines the term photochemotherapy. The wavelength of light used is determined by the activation spectrum of the chemical agent. The interaction of the light and chemical produces a synergistic effect. The clinical, pharmacologic, and toxic effects of the Goeckerman regimen, PUVA therapy, chemoporphyrin therapy, and the treatment of hyperbilirubinemia are reviewed.

The Goeckerman Regimen

The application of crude coal tar (CCT) followed the next day by ultraviolet light was first used at the Mayo Clinic by Goeckerman[1] in 1925. Since it was well known that sunlight generally improved psoriasis, it was Goeckerman's idea to potentiate this effect. The photosensitizing chemical crude coal tar (CCT) was used only because of a delay in obtaining hematoporphyrin. The light source was a hot quartz mercury vapor lamp emitting a discontinuous spectra of UVA and UVB.[1, 2] Numerous studies have confirmed the benefit of tar and light.[3–5]

CCT preparations have the disadvantage of being messy and odoriferous. In addition, hospitalization was required to administer a Goeckerman regimen properly. The development of psoriasis day care centers has allowed the use of the Goeckerman regimen on an outpatient basis. Cram and Menter cleared severe psoriasis (at least 90% improvement) in all 300 subjects treated with the Goeckerman regimen in an average of 3 weeks.[6] Even more impressive was the finding that 90% remained clear for a minimum of 8 months, and 73% remained clear for 1 year or longer following discharge from the day care center. The patients were encouraged to use liquor carbonis detergents (LCD) in a moisturizing base and UV lamps or sunlight at home. The success of these programs has encouraged the development of several of these centers, which offer lower-cost outpatient hospital care to the patient and insurance companies.

The main photosensitizing compounds found in coal tar are anthracene, methylanthracene, pyrene, fluoranthrene, and benzpyrene.[7] However, CCT is a mixture of more than 10,000 components and the chemical composition or biological activity of active ingredients has not been adequately characterized. Nonetheless, refined tar preparations have been developed which are more cosmetically acceptable and, therefore, more practical for home use. Studies have claimed effectiveness for these refined products.[8, 9] Variables in these studies make their evaluation difficult. The benefits of coal tar in psoriasis appear to be due to this agent's ability to cross-link with DNA, thus affecting DNA synthesis. It is interesting that the action spectrum for delayed erythema (phototoxic effect) of tar-treated skin is in the UVA, not the UVB, spectrum.[10]

Recent information has lead investigators to question the need for tar in the various modified Goeckerman regimens even though the treatment had been the mainstay of psoriasis for more than 50 years. Psoriasis has been cleared as effectively using plain petrolatum and ultraviolet light in the 280–400-nm range.[11, 12] The petrolatum was used to hydrate the scale and make it more transparent to the UV light. The role of coal tar in phototherapy of psoriasis has been defined better by Lowe and co-workers.[13] Their study compared CCT with plain petrolatum using erythemogenic and suberythemogenic doses of UVB. They found that CCT in petrolatum was superior to petrolatum alone only if suberythemogenic doses of UVB were used. The same finding occurred when a tar extract was compared to the oil base. The tar extract compared favorably to the crude tar. These regimens were repeated in the hairless mouse model to study the suppression of DNA synthesis. Erythemogenic UVB produced maximal inhibition of DNA synthesis, whereas suberythemogenic UVB produced submaximal suppression that was augmented by coal tar but not the vehicle.[13]

In spite of these studies questioning the need for coal tar and proposing newer cosmetically acceptable therapies, modified Goeckerman programs remain in wide use and have become the standard for comparing therapies for psoriasis. The regimen's popularity remains because of its effectiveness and long-term safety record. Contact dermatitis is rare and systemic toxicity does not occur. Even the potential for carcinogenesis appears to be minimal, since long-term human studies have not clearly demonstrated an increased incidence of skin cancer.[14–16] Unfortunately, these safety data are derived only from retrospective studies. Additionally, these data were derived before the use of current fluorescent lamps which emit principally UVB (peak 300 nm) and the accessibility of outpatient treatment centers and high-output home units. These factors will greatly increase the exposure of psoriatic patients to damaging UVB irradiation and substantially increase their risk of skin cancer. Additional concerns come from a recent study of blood extracts in volunteers who applied CCT for 8 hr on 2 consecutive days. This study documented absorption of phenanthrene, anthracene, pyrene, and fluoranthene, which have been shown to be tumor initiators or cocarcinogens. Therefore, continued prospective monitoring of psoriatic patients treated with the Goeckerman regimen is needed.[17]

PUVA

A decade ago it was found that psoralens given orally with longwave ultraviolet-A light (UVA) cleared psoriasis in most patients.[18] This very successful, cosmetically acceptable form of therapy was easily administered on an outpatient basis. With the development of high-output UVA lamps and devices to deliver correct doses of UVA accurately, the combination of *p*soralen and *UVA* (PUVA) therapy became popular. The FDA approved the use of PUVA for debilitating resistant cases of psoriasis after monitoring studies of more than 1,300 such patients at 16 medical centers in the United States. In these clinical trials, 88% of the patients were cleared of their symptoms, treatment failed in 3%, and only 10% had to drop out because of complications (3% dropped out for unrelated reasons).[19] Complete clearing usually requires 10–30 treatments, however, partial relief occurs after just 5 or 6 treatments, which encourages the patient to continue. Remissions are variable but seem to be of shorter duration than those reported with the Goeckerman regimen.[20] Perhaps this is due to the severity of the psoriasis: PUVA therapy is usually reserved for patients whose disease fails to respond to tar or topical steroid treatment.

Photochemotherapy with orally administered 8-methoxypsoralen has been used successfully to treat cutaneous T cell lymphoma (mycosis fungoides). Long-lasting remissions have been produced in early stages of this lymphoma.[21–23] "Parapsoriasis," a chronic condition of the skin characterized by scaling atrophic plaques and often seen as a precursor of mycosis fungoides, has also improved in most patients treated with PUVA.[24] Lymphomatoid papulosis, a pseudolymphoma of the skin that only rarely becomes a malignant T cell lymphoma, has been ameliorated by PUVA.[25]

Diseases that involve known abnormalities of the immune system respond favorably to PUVA. Atopic eczema,[26] alopecia areata,[27] and vitiligo[28] have been treated successfully with PUVA photochemotherapy. Other skin diseases that involve possible immune dysfunction—lichen planus,[29] polymorphous light eruption,[30] and solar urticaria[31]—have improved with PUVA therapy. An increasing number of studies suggest an immunologic basis to the pathogenesis of psoriasis, which is reversed by PUVA.[32] A decrease in T lymphocytes and an interference with the immune mechanisms in psoriasis occurs following PUVA therapy.[33] The increased chemotactic activity of polymorphonuclear leukocytes in psoriasis can be inhibited by PUVA.[34] Patients with psoriasis have abnormalities in their cell-mediated immune responses, and their serum contains a factor that inhibits normal formation of T lymphocyte E rosettes. This inhibition was normalized by PUVA treatment.[35] Other T-cell functions have been shown to be defective in patients with psoriasis. PUVA has produced decreased DNA synthesis of lymphocytes *in vivo* and *in vitro*.[36] It has also decreased PHA stimulation of lymphocytes.[36] The precise role of these effects on lymphocytes in modulating psoriasis is still unknown.

The photosensitization reaction produced by psoralens is similar in many respects to that induced by any photoreactive chemical, that is, photoexcitation to singlet and triplet states leading to the generation of free radicals and singlet oxygen. These reactive substances can damage cells. In addition, when psoralens are photoactivated, they bind covalently to pyrimidine bases of DNA, forming cycloadducts and monoadducts. They can also cause interstrand cross-linking of DNA,[37, 38] These reactions with DNA are thought to be responsible for the inhibition of DNA synthesis and results in the dampening of cell turnover, resulting in the therapeutic benefits in psoriasis. The pathologic changes following PUVA therapy of psoriasis support the concept that DNA damage occurs. Epidermal cell dystrophy is present and the nuclei and nucleoli of keratinocytes are altered, suggesting alterations in the chromosomes.[39] This cytotoxic mechanism might also explain the inhibition of immune responses in the skin.

The pharmacokinetics and animal toxicology of PUVA have been extensively studied and reviewed by Wulf.[36] In rats, 8-methoxypsoralen and its metabolites accumulate in the centrilobular area of the liver, the corticomedullary region in the kidney, in the adrenal cortex, in hair follicles, and in the pigmented layer of the retina. Cataract and pigmented retinopathy have been seen in a variety of animal studies but not with therapeutic doses.[36] Liver and kidney toxicity have been minimal.[36] Skin cancer has been induced in several animal experiments but not after physiological doses.[36, 37]

It has been shown in humans that more than 95% of orally administered 8-methoxypsoralen is absorbed by the gastrointestinal tract and that blood levels peak between 1½ and 3 hr after ingestion of the drug.[38] Adequate renal function is important in the excretion of the drug. More than 70% of the drug is eliminated in 8 hr and 90% of metabolites can be recovered in the urine between 12 and 24 hr.[38] Caution must be exercised in treating patients with hepatic insufficiency, because hepatic biotransformation is necessary for urinary excretion.[38] Knowledge of the appropriate toxicology and pharmacokinetics allows one to administer therapy intelligently. Renal and hepatic functions are evaluated before therapy and periodically during treatment if long-term therapy is required. Patients are given the drug 2 hr before exposure to UVA. They are then instructed to avoid UV exposure for 24 hr following the dose of psoralen and to wear UV-shielding glasses during the day of the treatment to protect the eyes and prevent cataracts. Routine examination of the eye and skin is required. Subjects with fair skin should receive lower doses of light and should be monitored carefully for the development of skin cancers.

The common side effects, which occur in at least 10–20% of the PUVA-treated patients, are nausea and pruritus. The pruritus is of two types: a more common mild itch that is nonspecific and a rarer, severe itch that may necessitate discontinuation of therapy. The latter itch has also been described as a pain or burning sensation that comes in bouts and is usually localized and unresponsive to therapy.[40, 41] Concern about possible long-term side effects is the major deterrent to the use of PUVA. These side effects include ocular damage and carcinoma of the skin. Although animal studies have raised concern about ocular damage with the use of PUVA, no ocular abnormalities were attributed to PUVA therapy in an extensive ophthalmologic study of 46 patients treated with PUVA for 5–6½ years.[42] The absence of ophthalmologic complications may well be due to the precautions noted above. It is still necessary to recommend that patients wear UVA-blocking glasses for at least 12 hr after psoralen ingestion, since free 8-MOP has been detected in the human lens up to 10 hr after oral ingestion.[43]

The risk of skin cancer in patients treated with PUVA has been debated in the literature. Results from the United States cooperative study strongly suggest an increased risk of skin cancer, especialy squamous cell carcinoma.[44] It is not surprising that this risk was increased further if the patient was fair-skinned and had a prior history of exposure to ionizing radiation.[45] The most convincing evidence was the reversal of the squamous cell to basal cell carcinoma ratio and the increased appearance of tumors in normally covered areas of the body.[44] In a continued prospective study of 1,380 patients with psoriasis, Stern and co-workers found a 12.8-fold increased risk of squamous cell carcinoma in patients receiving high-dose compared to low-dose PUVA.[45] There was no dose-related change in the occurrence of basal cell carcinomas.[45] In smaller European studies of PUVA patients, the increased number of skin cancers was not attributed directly to the use of PUVA but to other potentially carcinogenic treatment, especially ionizing radiation.[46–48] In another large cooperative study of 631 patients evaluated for 4 years after PUVA therapy, only 10 patients developed skin cancers.[49] There were 10 basal cell carcinomas and 3 squamous cell carcinomas. The number and distribution of these cancers do not suggest any association with PUVA therapy. In addition, two-thirds of the patients had been exposed to previous ionizing radiation.[49] A study of 2,013 patients, with various skin diseases treated with PUVA therapy over a 10-year period, did not show an increase in the spontaneous rate of skin cancer.[50] In assessing risk, it must be remembered that the majority of basal cell and squamous cell tumors are easily treated, while severe psoriasis is a physically and socially disabling disease.

Other changes of note have been documented in subjects receiving long-term PUVA therapy, including atypical pigmented lentigines (freckles) wrinkling, atrophy, and telangiectasias of the skin.[47, 51] These changes associated with aging, may be of more concern to patients than the risk of skin cancer.

Topical psoralen therapy has been used to avoid the side effects associated with systemic administration. A variety of methods have been tried, from total bath immersion to very limited application of alcohol or ointment preparations. Concentrations have varied from 0.00005 to 1%, and application to exposure time has varied from immediate to 2 hr. Psoralens appear to penetrate rapidly but the penetration is variable and unpredictable.[52] Topical psoralen therapy is tedious to administer and may often result in blistering because of a narrow margin of safety.[53] Hyperpigmentation often occurs around the treated sites.[53] Topical therapy does avoid the side effects of nausea and the need to protect the eyes and the untreated skin from UV light for 24 hr. It is especially useful in the treatment of localized areas of vitiligo where one wants to avoid hyperpigmenting the normal skin.

Isotretinoin has been given in combination with PUVA to lower the total dose of PUVA necessary to clear psoriasis. Since this has worked successfully and isotretinoin alone has not been very effective in psoriasis, its action may represent a synergistic effect.[54] This may turn out to be a helpful adjunct to decrease the time and side effects of PUVA therapy.

Studies[55] of specific bands of UVR have shown that the optimal action spectrum for phototherapy of psoriasis is around 313 nm. This is above the peak emission of presently used fluorescent UVB units. Perhaps with the development of more sophisticated light sources, which deliver limited but ideal wavelengths, tar or psoralens to augment UV light will be unnecessary.

Phototherapy of Neonatal Hyperbilirubinemia

This treatment gained widespread use following Cremer's observation that serum bilirubin concentrations fell more rapidly in infants closest to the windows than those in darker areas of the nursery.[56] Now phototherapy represents a major method for the prevention and treatment of bilirubin toxicity in the newborn. One can determine if an infant may fit the guidelines for phototherapy after a thorough investigation into the cause of its jaundice. A 1974 survey of American hospitals found that 57% used phototherapy and that an estimated 2.5% of the neonates were exposed to light treatments.[57]

The efficacy of phototherapy has been substantiated in a randomized, controlled trial supported by the National Institute of Child Health and Human Development (NICHHD).[58] The study included 1,339 infants with 672 assigned to the phototherapy group, and 667 serving as controls. Infants were classifed and treated according to birth weight. Phototherapy was found to be effective in the prevention of hyperbilirubinemia and in decreasing the number of exchange transfusions in babies in all weight groups when compared to controls.[59] This was true even in low-birth-weight infants (< 2,000 g) with hemolysis. In the group weighing 2,500 g or greater and suffering from Coombs'-positive hemolytic disease, continuous phototherapy did not decrease the need for exchange transfusion.[60] Light therapy was found to be most efficacious in the first 24–48 hr of use and especially when initiated in the first day of life. Fluorescent bulbs emitting maximum energy at 410–460 nm are most effective for the elimination of bilirubin.

Phototherapy lowers serum bilirubin concentrations because it generates more polar, readily excretable bilirubin compounds. When bilirubin is exposed to light, two chemical changes take place: photoisomerization and photooxidation.[61]

Photoisomerization occurs when the bilirubin molecule absorbs a photon of light. These isomers have more polar functions exposed on their surface than their preferred conformation, bilirubin. These more polar soluble compounds migrate into the circulatory system where they become bound to albumin and are extracted from the blood into the hepatocytes. Here the isomers, which do not require conjugation for excretion, are excreted into the bile.[62]

Photooxidation involves the production of an excited form of bilirubin. Triplet bilirubin transfers its energy to molecular oxygen to create singlet oxygen. The latter reacts with bilirubin, producing lower-molecular-weight, more polar products,[63] Given the ease with which bilirubin undergoes photoisomerization *in vitro*, this reaction is thought to predominate over the slower and less efficient photooxidation process *in vivo*.[63]

Phototherapy presumably affects the bilirubin in the extravascular compartment within 2 mm of the skin surface.[64] The lowering of serum bilirubin is thought to be a secondary effect of treatment in which there is an efflux of bilirubin from the circulation into the vacated skin sites.

The phototoxic effects of this therapy can be acute or long-term. Continuing clinical studies are investigating the latter. The most commonly reported acute side effects include: rash, bronze baby syndrome (in infants with obstructive jaundice), transient loose green stools, lethargy, abdominal distention, and accelerated tanning of black infants.[61, 64]

Metabolic effects include: altered oxygen consumption,[65] more rapid respiratory rate,[66] increased blood flow,[67] and increased water loss.[66, 68] These alterations may be explained in part by thermal effects of light including both increased environmental and skin surface temperatures.[65] Phototherapy also appears to decrease gut transit time.[69, 70] The increased biliary excretion of unconjugated bilirubin apparently inhibits intestinal lactase. The resultant increased number of stools and diarrhea decreases absorption of lactose, water, and possibly other nutrients. The consequences in terms of early growth and development are unclear. However, the early deficit in weight gain in light-treated infants is quickly made up over the subsequent 2 or 3 weeks.[71]

Biochemical effects of unknown clinical and biological significance include: altered quantity and nature of tryptophan metabolites excreted in urine,[72] increased plasma levels of growth hormone, luteinizing, and follicle-stimulating hormones (in female patients only),[74, 75] increased platelet turnover,[76] and decreased immunoreactive prostaglandin-A values in premature infants.[77] The eye shields necessary during treatment deprive the infant of much-needed visual stimulation during a critical period of development.

Photoproduct toxicity does not seem to be a consequence of the photobilirubins, since they are rapidly excreted. Singlet oxygen is a documented cytotoxic agent *in vitro*. However, there is no evidence of this photoproduct causing photodynamic injury in patients receiving phototherapy.[61]

Klaus and Fanaroff[78] recommend that the following guidelines be used when an infant is placed without clothing under the lamp in a closed incubator: (1) monitor the temperature of the infant every 4 hr; (2) weigh the infant every 8 or 12 hr and watch for significant weight loss secondary to increased evaporative water loss and diarrhea; (3) shield the eyes and gonads from the light; (4) change the infant's position every 6 hr; (5) measure serum bilirubin levels every 6–8 hr; (6) discontinue phototherapy whenever the mother visits her infant and remove the eye shields to allow natural interaction between the mother and her infant; (7) monitor the bilirubin levels every 8–12 hr for a day after the phototherapy is discontinued to detect any rebound rise in serum bilirubin concentration.

Chemoporphyrin Therapy

Hematoporphyrin derivative (HpD) is an investigational photosensitizing compound that has been demonstrated to be of diagnostic and therapeutic benefit in neoplasia.[79-82] HpD is an ideally suited agent for three reasons:[83] (1) it preferentially localizes in anaplastic and cancerous cells, (2) it emits a characteristic salmon-red fluorescence upon exposure to light of appropriate wavelengths, and (3) it is capable of mediating a cytotoxic photochemical reaction when excited by light.

The factors accounting for the selective affinity of HpD for certain cells are unknown,[82] but this property makes it a useful tumor marker, especially in clinically occult lesions. One clinical application of this property is in solving the urologic contradiction of positive results of urine cytologic study and negative findings on routine cytoscopic exam. HpD would contribute to the localization of carcinoma in situ of the bladder using fluorescent endoscopy.

When HpD is irradiated with light of the appropriate wavelengths, cytotoxic compounds are generated by the transfer of energy from an excited porphyrin to oxygen. The major cytotoxic agent is believed to be singlet oxygen.

Photoradiation therapy using HpD was first popularized by Dougherty and colleagues[79] in 1978. Since then, it has been used to study lesions of the skin, breast, bladder, brain, and esophagus; however, most of the investigations have been in pulmonary lesions. The usual technique consists of the intravenous administration of HpD followed by laser phototherapy administered through a fiberoptic endoscope.[81]

This procedure has certain limitations. Only lesions accessible by the fiberoptic bronchoscope can be treated, and only lesions less than 1.0 cm deep can be treated because of limitations in the penetration of the light.

Photoradiation therapy has proved to be of value in patients with in situ and early invasive lesions who are poor surgical risks. It is also of potential use in the treatment of lesions that traditionally preclude an operative procedure, such as carcinomas invading the trachea or carina.[84] Pulmonary complications include localized tissue edema, heavy sputum production, and hemoptysis. Photosensitivity may also complicate this therapy.

References

1. W. H. Goeckerman, Treatment of psoriasis, *Northwest. Med.* *24*:229-231, 1925.
2. S. A. Muller and H. O. Perry, The Goeckerman treatment in psoriasis. Six decades of experience at the Mayo Clinic, *Cutis 34*:265, 1984.
3. H. O. Perry, C. W. Soderstrom, and R. W. Schultze, The Goeckerman treatment of psoriasis, *Arch. Derm. 98*:178-182, 1968.
4. A. R. Marsico, W. H. Eaglstein, and G. O. Weinstein, Ultraviolet light and tar in the Goeckerman treatment of psoriasis, *Arch. Derm. 112*:1249–1250, 1976.
5. E. Epstein, Photosensitization in the treatment of psoriasis, *J. Invest. Derm. 8*:203–207, 1947.
6. A. Menter and D. L. Cram, The Goeckerman regimen in two day care centers, *J. Amer. Acad. Derm. 9*:59, 1983.
7. W. Burckhardt, Zurfrage der Photosensibilisierunder, Wirjung des Teers, *Schweiz. Med. Wochenschr. 83*, 1939.
8. W. Jordan, A. Clarke, and R. Hale, Long term modified Goeckerman regimen for psoriasis using an ultraviolet B light source in the home, *J. Amer. Acad. Derm. 4*:584–591, 1981.
9. D. L. Cram, Psoriasis treatment with a tar gel, *Cutis 17*:1197–1198, 1976.
10. L. Tanenbaum, J. A. Parrish, M. A. Pathak, R. R. Anderson, and T. B. Fitzpatrick, Tar phototoxicity and phototherapy for psoriasis, *Arch. Derm. 111*:467–470, 1975.
11. M. J. Levine, H. A. D. White, and J. A. Parrish, Components of the Goeckerman regimen, *J. Invest. Derm. 73*:170–173, 1979.
12. M. J. Levine and J. A. Parrish, Outpatient phototherapy of psoriasis, *Arch. Derm. 116*:552, 1980.
13. N. J. Lowe, M. S. Wortzman, J. Breeding, H. Koudsi, and L. Taylor, Coal tar phototherapy for psoriasis reevaluated: Erythemogenic versus suberythemogenic ultraviolet with a tar extract in oil and crude coal tar, *J. Amer. Acad. Derm. 8*:781–789, 1983.
14. T. F. Anderson, T. P. Waldinger, and J. J. Vorrhees, UV-B phototherapy: An overview, *Arch. Derm. 120*:1502–1507, 1980.
15. M. R. Pittelkow, H. O. Perry, S. A. Muller, W. Z. Maugham, and P. C. O'Brien, Skin cancer in patients with psoriasis treated with coal tar, *Arch. Derm. 117*:465–468, 1981.
16. R. S. Stern, S. Zierler, and J. A. Parrish, Skin carcinoma in patients with psoriasis treated with topical tar and artificial radiation, *Lancet 1*:732–735, 1980.
17. J. S. Storer, I. DeLeon, L. E. Millikan, J. L. Laseter, and C. Griffing, Human absorption of crude coal tar products, *Arch. Derm. 120*:874–877, 1984.
18. J. A. Parrish, T. B. Fitzpatrick, L. Tanenbaum, and M. Pathak, Photochemotherapy of psoriasis with oral methoxsalen and longwave ultraviolet light, *New Engl. J. Med. 291*:1207. 1974.
19. J. W. Melski, L. Tanenbaum, J. A. Parrish, T. B. Fitzpatrick, and H. L. Bleich, Oral methoxsalen photochemotherapy for the treatment of psoriasis: A cooperative clinical trial, *J. Invest. Derm. 68*:328–335, 1977.

20. J. Beer, J. Hermans, A. Schothorst, and D. Suurmond, Comparison of phototherapy (UV-B) and photochemotherapy (PUVA) for clearing and maintenance therapy of psoriasis, *Arch. Derm. 120*:52, 1984.
21. J. A. Parrish, M. J. LeVine, and T. B. Fitzpatrick, Oral methoxsalen photochemotherapy of psoriasis and mycosis fungoides, *Int. J. Derm. 19*:379–386, 1980.
22. H. H. Roenigk, Photochemotherapy for mycosis fungoides, *Arch. Derm. 113*:1047, 1977.
23. H. Honigsmann, W. Brenner, W. Rauschmeier, K. Konrad, and K. Wolff, Photochemotherapy of cutaneous T-cell lymphoma, *J. Amer. Acad. Derm. 10*:238–245, 1984.
24. F. C. Powell, G. T. Spiegel, and S. A. Muller, Treatment of para-psoriasis and mycosis fungoides. The role of psoralens and long wave ultraviolet light PVA, *Mayo Clin. Proc. 59*:538–546, 1984.
25. G. L. Wantzin and K. Thomsen, PUVA treatment of lymphomatoid papulosis, *Brit. J. Derm. 107*:687–690, 1982.
26. W. L. Morison, J. A. Parrish, and T. B. Fitzpatrick, Oral psoralen photochemotherapy of topic eczema, *Brit. J. Derm. 98*:25, 1978.
27. A. Lassus, U. Kianto, E. Johansson, and T. Juvakoski, PUVA treatment of alopecia areata, *Dermatologica 161*:293–304, 1980.
28. J. A. Parrish, T. B. Fitzpatrick, C. Shea, and M. A. Pathak, Photochemotherapy of vitiligo, *Arch. Derm. 112*:1531, 1976.
29. J. P. Ortonne, J. Thivolet, and C. Sannwald, Oral photochemotheapy in the treatment of lichen planus, *Brit. J. Derm. 99*:77, 1978.
30. J. A. Parrish, M. J. LeVine, W. Morrison, E. Gonzalez, and T. B. Fitzpatrick, Comparison of PUVA and beta-carotene in the treatment of polymorphous light eruption, *Brit. J. Derm. 100*:187, 1979.
31. J. A. Parrish, K. F. Jaenicke, W. L. Morrison, K. Montaz, and C. Shea, Solar urticaria: Treatment with PUVA and mediator inhibitors, *Brit. J. Derm. 106*:575–580, 1982.
32. W. L. Morrison, J. A. Parrish, and J. H. Epstein, Photoimmunology, *Arch. Derm. 115*:330–355, 1979.
33. R. H. Curmane, E. Hamerlinck, and A. H. Siddiqui, Immunologic implications of PUVA therapy in psoriasis vulgaris, *Arch. Derm. Res. 265*:245–267, 1979.
34. N. Mizano, H. Enami, and K. Eraki, Effect of 8-methoxypsoralen plus UVA on psoriasis leukotactic factor, *J. Invest. Derm. 72*:64–66, 1979.
35. M. Haftek, W. Glinski, S. Jablonski, and S. Obalek, T lymphocyte E rosette function during photochemotherapy (PUVA) of psoriasis, *J. Invest. Derm. 72*:214–218, 1979.
36. H. C. Wulf, Aspects of toxicology of 8-methoxypsoralen and longwave ultraviolet light, *Acta Dermatovener. Suppl. 106*:1–42, 1982.
37. A. Bredberg, Genetic toxicity of psoralen and ultraviolet radiation in human cells, *Acta Dermatovener. Suppl. 104*:1–40, 1982.
38. M. Pathak, T. B. Fitzpatrick, and J. A. Parrish, Pharmacologic and molecular aspects of psoralen phototherapy, In *Psoriasis—Proceedings of the Second International Symposium*, E. M. Farber and A. J. Cox, Eds., Yorke Medical Books, New York, 1977, pp. 262–271.
39. K. Ree, A. S. Johnson, and T. Hovig, Ultrastructural studies on the effects of photoactivated 8-methyoxpsoralen, *Acta Pathol. Microbiol. Scandinav (A) 89*:81–90, 1981.
40. E. Tegner, Severe pain after PUVA treatment, *Acta Derm. Venereol. (Stockh.) 60*:187, 1980.
41. J. Miller, Severe pain following PUVA, *Acta Derm. Venereol. (Stockh) 60*:187, 1980.
42. L. Ronnerfalt, E. Lydahl, G. Wennersten, P. Jahnberg, and M. Thyresson-Hok, Ophthalmologic study of patients undergoing long term PUVA therapy, *Acta Derm. Venereol. 62*:501–505, 1982.
43. S. Lerman, J. Megaw, and I Willis, Potential ocular complications from PUVA therapy and their prevention, *J. Invest. Derm. 74*:197, 1980.
44. R. S. Stern, L. A. Thibodeau, R. A. Kleinerman, J. A. Parrish, and T. B. Fitzpatrick, Risk of cutaneous carcinoma in patients treated with oral methoxysalen photochemotherapy for psoriasis, *New Engl. J. Med. 300*:809–813, 1979.

45. R. S. Stern, N. Laird, J. Melski, J. A. Parrish, T. B. Fitzpatrick, and H. Bleich, Cutaneous squamous cell carcinoma in patients treated with PUVA, *New Engl. J. Med. 310*:1156–1161, 1984.
46. A. Lassus, T. Reunala, J. Idanpaa-Heikkila, T. Juvakoski, and O. Salo, PUVA treatment and skin cancer. A follow-up study, *Acta Derm. Venereol. 61*:141–145, 1981.
47. M. Rosa, G. Wennersten, and B. Langerholm, Longterm phototherapy for psoriasis: A histopathological and clinical follow-up study with special emphasis on tumor incidence and behavior of pigmented lesions, *Acta Dermatovener. 63*:215, 1983.
48. R. Lindskov, Skin carcinomas and treatment with photochemotherapy (PUVA), *Acta Dermatovener. 63*:223, 1983.
49. H. H. Roenigk, Jr. and W. A. Caro, Skin cancer in PUVA 48 cooperative study, J. Amer. Acad. Derm. 4:319–324, 1981.
50. G. Stuttgen, The risk of photochemotherapy, *Int. J. Derm. 21*:198–202, 1982.
51. A. R. Rhodes, T. J. Harrist, and K. Montaz, The PUVA induced pigmented macule. A lentiginous proliferation of large, sometimes cytologically atypical melanocytes, *J. Amer. Acad. Derm. 9*:47–58, 1983.
52. K. Dano, T. Horio, M. Ozaki, and S. Imamura, Topical 8 methoxypsoralen photochemotherapy of psoriasis, clinical study, *Brit. J. Derm. 108*:519–524, 1983.
53. J. A. Petrozzi and A. M. Kligman, Topical psoralen photochemotherapy, In *Psoriasis, Proceedings of the Second International Symposium*, E. M. Farber and A. J. Cox, Eds., Yorke Medical Books, New York, 1977, pp. 285–290.
54. H. Honigsman and K. Wolff, Isotretinoin PUVA for psoriasis, *Lancet 1*:236, 1983.
55. J. A. Parrish and K. F. Jaenicke, Action spectrum for phototherapy of psoriasis, *J. Invest. Derm. 76*:359–362, 1981.
56. R. J. Cremer, Influence of light on the hyperbilirubinemia of infants, *Lancet 1*:1094, 1958.
57. R. M. Albrecht, Phototherapy for neonatal hyperbilirubinemia: A survey of U.S. hospitals in 1974, In Symposium on biological effects and measurement of light sources, DHHS pub. No. (FDA) 81-8156, 1981, 55-68.
58. National Institute of Child Health and Human Development, *Pediatrics 75*(2) Suppl.:385–441, 1985.
59. A. K. Brown, M. H. Kim, P. Y. K. Wu, and D. A. Byrla, Efficacy of phototherapy in prevention and management of neonatal hyperbilirubinemia, *Pediatrics 75*(2) Suppl.:393–400, 1985.
60. H. M. Maurer, B. V. Kirkpatrick, N. B. McWilliams, D. A. Draper, and D. A. Byrla, Phototherapy for hyperbilirubinemia of hemolytic disease of the newborn, *Pediatrics 75*(2) Suppl.:407–412, 1985.
61. A. K. Brown and A. F. McDonugh, Phototherapy for neonatal hyperbilirubinemia: Efficacy, mechanism and toxicity, *Adv. Pediatr. 27*:341–390, 1980.
62. P. Y. K. Wu, J. E. Housman, B. V. Kirkpatrick, N. B. White, M. Phil, and D. B. Byrla, Metabolic aspects of phototherapy, *Pediatrics 75*(2) Suppl:427–433, 1985.
63. T. P. Vogl, Phototherapy of neonatal hyperbilirubinemia; bilirubin in unexposed areas of the skin, *J. Pediatr. 85*:707, 1974.
64. D. A. Byrla, Development, design and sample composition, *Pediatrics 75*(2) Suppl.:387–392, 1985.
65. O. R. C. Sonales, Effect of phototherapy on thermal environment of the newborn, *Arch. Dis. Child. 53*:172, 1978.
66. W. Oh and H. Karecki, Phototherapy and insensible water loss in the newborn infant, *Am. J. Dis. Child. 124*:230, 1972.
67. P. Y. K. Wu, W. H. Wong, J. E. Hodgman, and N. Levan, Changes in blood flow in the skin and muscle with phototherapy, *Pediatr. Res. 8*:257, 1974.
68. P. Y. K. Wu and J. E. Hodgman, Insensible water loss in preterm infants: Changes with postnatal development and non-ionizing radiant energy, *Pediatrics 54*:704, 1974.
69. F. F. Rubatelli and G. Largojolli, Effect of light exposure on gut transit time in jaundiced newborn, *Acta Pediatr. Scand. 62*:146, 1973.
70. P. Y. K. Wu and A. Moosa, Effect of phototherapy on nitrogen and electrolyte levels and water balance in jaundiced preterm infants, *Pediatrics 61*:193, 1978.

71. P. Y. K. Wu, R. C. Lim, J. E. Hodgman, M. J. Kokosky, and A. Teberg, Effect of phototherapy in preterm infants on growth in the neonatal period, *J. Pediatr.* *85*:563, 1974.
72. F. F. Rubaltelli, G. Allegri, C. Costa, and A. DeAntoni, Urinary excretion of tyrptophan metabolites during phototherapy, *J. Pediatr.* *85*:865, 1974.
73. K. E. V. Muhlendahl and L. Ballowicz, Growth hormone and cortisol in neonates during phototherapy, *Z. Kinderheilkd.* *119*:53, 1975.
74. C. Dacou-Voutetakis, D. Anagnostakis, and D. Matsaniotis, Effect of prolonged illumination on concentrations of leutinizing hormone in human infants, *Science 199*:1229, 1978.
75. B. J. Lemaitre, Increased gonadotrophin levels in newborn premature infants treated by phototherapy, *J. Steroid Biochem.* *10*:35, 1979.
76. H. M. Maurer, M. Fratkin, N. B. McWilliams, B. Kirkpatrick, O. Draper, J. D. Higgins, and C. R. Hunter, Effects of phototherapy on platelet counts in low birth weight infants and on platelet production and life span in rabbits, *Pediatrics 52*:506, 1976.
77. C. E. Aplin, B. H. Brouhand, R. J. Cunningham, and C. J. Richardson, Phototherapy and plasma immunoreactive prostaglandin A values, *Am. J. Dis. Child.* *133*:625, 1979.
78. M. H. Klaus and A. A. Fanaroff, *Care of the High Risk Neonate*, W. B. Saunders, Philadelphia, 1979.
79. T. J. Dougherty, J. E. Kaufman, A. Goldfarb, K. R. Weishaupt, D. Boyle, and A. Mittleman, Photoradiation therapy for the treatment of malignant tumors, *Cancer Res.* *38*:2628, 1978.
80. Y. Hyata, H. Kato, C. Konaka, J. Ono, and N. Talizman, Hematoporphyrin derivative and laser photoradiation in the treatment of lung cancer, *Chest 81*:269, 1982.
81. D. Cortese and J. H. Kinsey, Endoscopic management of lung carcinoma with hematoporphyrin derivative phototherapy, *Mayo Clin. Proc.* *57*:543, 1982.
82. R. C. Benson, Detection and localization of in situ carcinoma of the bladder with hematoporphyrin derivative, *Mayo Clin Proc.* *57*:548–555, 1982.
83. J. H. Kinsey, Photodynamic effect of HpD as a function of optical spectrum and incident energy density, *Cancer Res.* *41*:5020, 1981.
83. T. J. Dougherty, Editorial, Photoradiation therapy for bronchogenic cancer, *Chest 81*:265–266, 1982.

7

PHOTOEFFECTS ON THE EYE

PIERRETTE DAYHAW-BARKER, Ph.D.
FELIX M. BARKER II, O.D.
The Pennsylvania College of Optometry
Philadelphia, Pennsylvania

Abstract

A number of conferences and workshops in recent years have highlighted the growing interests of vision researchers in the role played by radiation on the integrity of ocular tissues. We review the clinical manifestations and discuss the basic research presently being carried out, with special emphasis on the photochemical events believed to contribute to lenticular and retinal alterations. The clinical battery of tests presently available is discussed in relation to pathologic and toxicologic entities encountered in practice. We hope that greater attention will be given to the early detection of photorelated visual dysfunctions.

Introduction

For the eye to fulfill its main function as the sensory organ of vision, radiation, more specifically the visible spectrum, must penetrate the anterior ocular tissues and be transmitted to the retina for absorption. The optical spectrum (i.e., the ultraviolet, visible and near infrared wavelengths) also represents a hazard to the integrity of ocular tissues.

This article will review the particular transmittance characteristics of the ocular tissues, the process of photosensitization by endogenous as well as exogenous photosensitizers, the deleterious photoeffects in ocular tissues, as well as the clinical methods used to detect such effects in patients at risk.

Ocular Transmittance

Figure 1 shows the gross anatomy of the eye. Processing of light for vision is accomplished by the optical tissues—cornea, aqueous humor, lens, and vitreous humor—which receive light in that order and focus it on the neural retina.

The ability of radiant energy to damage the ocular tissues is largely determined by the spectral transmittance/absorbance properties of the individual tissues. Thus the optical tissues of the eye that precede one another in the path of the incident

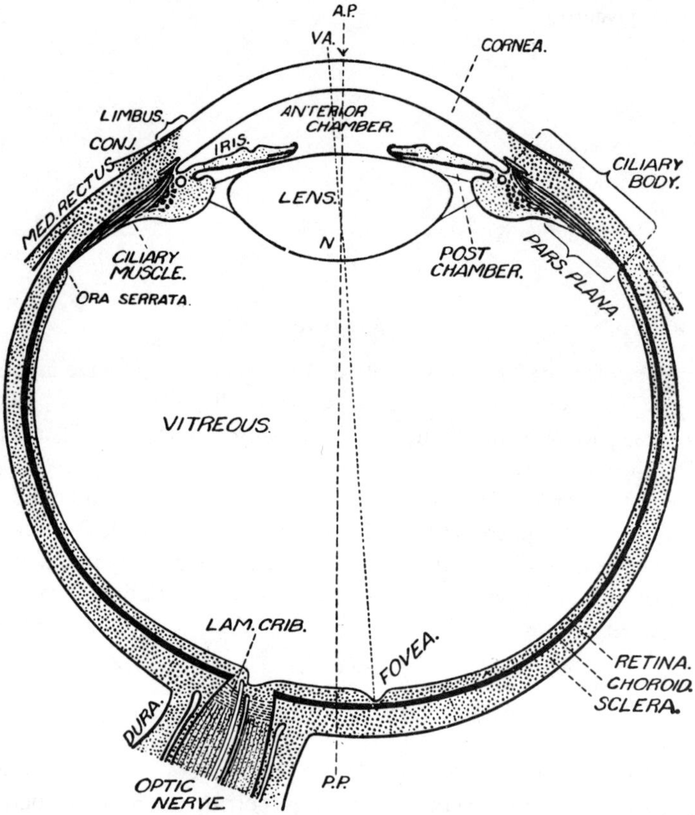

Figure 1. Schematic of the eye (from E. Wolff, *Anatomy of the Eye and Orbit*, seventh edition, revised by R. Warwick, W. B. Saunders, Philadelphia, 1976.)

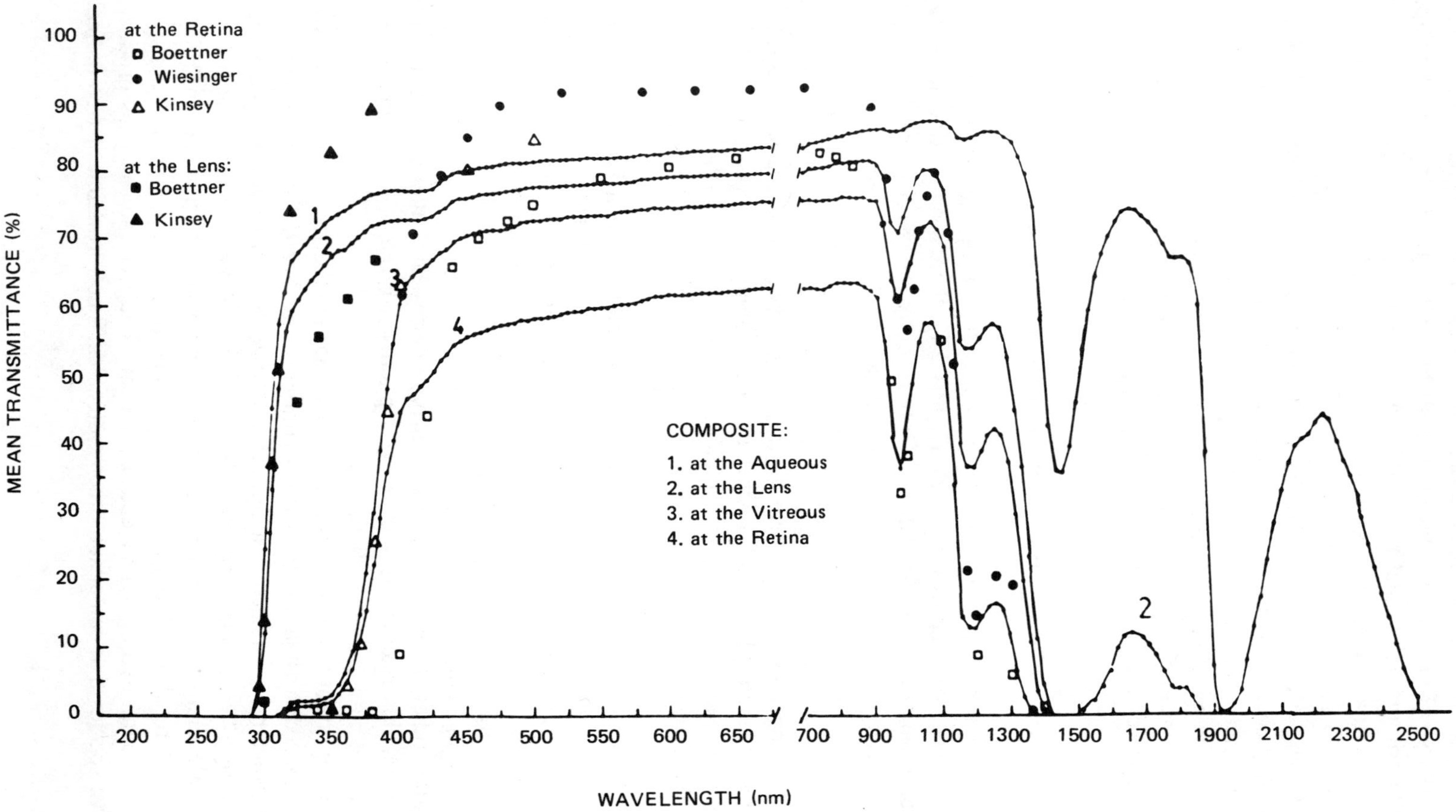

Figure 2. Spectral transmittance charcteristics of the human (■) and the rabbit (•, ▲) eye (from F. Barker, The transmittance of the electromagnetic spectrum from 2500 nm through the optical tissues of the eye of the pigmented rabbit, Thesis, Univ. of Houston, 1979).

energy offer filtration effects that are important in restricting exposure of the inner tissues to energy. For example, the cornea absorbs all of the energy incident upon it from 295 nm and below and transmits above 295 nm. In considering the photoeffects on lenticular tissue, one is therefore concerned primarily with wavelengths greater than 295 nm.

The transmittance of ocular tissues is shown in Figure 2. The transmittances of the cornea, aqueous, and vitreous are approximately the same regardless of species, whereas lens transmittance is different and specific for each species. This difference is significant since it alters the nature of incident energy on the retina. There is also a tendency for lens transmittance to decrease as a function of increasing age, especially the blue region of the spectrum.[1]

Photochemistry and Photosensitization

It has long been known that exposure to solar radiation can, under certain conditions, result in acute ocular disturbances. Historically, there is ample documentation of solar retinopathy and snow-blindness. Only recently, however, have more subtle effects been recognized, primarily with the development of cataracts and alterations in the photoreceptors and retinal pigment epithelium of the retina. These are presently being investigated in a number of laboratories and clinical settings.

Some basic concepts of photochemistry are necessary for an understanding of the mechanisms through which such disturbances are thought to occur. To initiate a photochemical reaction, a specific wavelength of radiation must be incident upon a chromophore and be absorbed by it. This initiates a chain of events including excitation into the singlet state 1_S (usually short-lived) or via intersystems cross-over to a triplet state. The triplet state (3_S) has a longer half-life than 1_S and can potentially interact longer with molecules in its environment. Return to the ground state can involve fluorescence, phosphorescence, or internal conversion (Fig. 3). Failure to return to the ground state because of the formation of photoproducts is, by definition, a photochemical event.[2] If the photoproduct involves intramolecular reorganization of the original chromophore, the reaction is said to be photochemical in nature; if molecules other than the original chromophores are altered and become photoproducts, this reaction is said to be a photosensitized one, and the original chromophore is now called a photosensitizer.[3]

It is also important to our understanding of ocular photoeffects that the following fact be recognized: a photoproduct from an initial reaction can, if it has the particular molecular structure, itself become a chromophore/photosensitizer in susbsequent reactions, initiating a "cascadelike sequence." A number of hypotheses about the development of senile cataracts are based on the assumption of such a sequence.

Lastly, let us recall that Type I reactions produce free radicals by direct substrate interaction through hydrogen or electron transfer. The free radicals thus produced are reactive species that can attack other molecules. The formation of free radicals

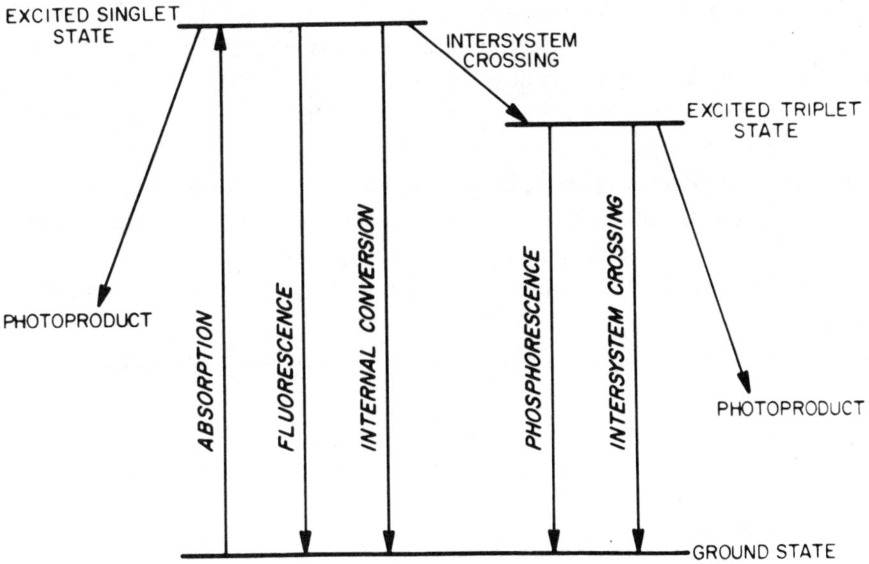

Figure 3. Schematic of processes available to excited states (from I. E. Kochevar, Basic concepts in photobiology, In *Photoimmunology*, J. A. Parrish, M. K. Kripke, and W. L. Morison, Eds., Plenum Press, New York, 1983).

does not always require radiation,[4] although radiation is one of the main ways by which free radicals are generated. A Type II (photodynamic) reaction involves 3_S excitation of ground state oxygen to $1O_2$. Singlet oxygen is recognized as dangerous to many molecules because it is both highly reactive and has a relatively long half-life compared to other singlet states.

The formation of the free radicals of oxygen, as well as singlet oxygen, is now known to be involved in many disease processes, including lipid peroxidation, a process that destroys cell structure and function. Formation of free radicals can result in alteration in nucleic acids and polymerization as well as depolymerization of proteins, which also incapacitates cell function.

The chemicals that act as chromophore/photosensitizer share a conjugated double-bond structure and are often tricyclic or heterocyclic compounds. A large number of endogenous chromophores are present in ocular tissues. In the lens, the nucleic acids and the aromatic amino acids, especially tryptophan, are of particular significance. In the retina, a number of molecules in addition to rhodopsin and cone pigments can interact, the most notable of which is melanin.[5]

While the higher-energy photons (high-frequency, low-wavelengths) are extremely efficient in producing electronic excitation, longer wavelengths, such as infrared, which contribute mainly vibrational energy, can exacerbate a coincident photic reaction.[5]

Thus a number of researchers now believe, for example, that senile cataracts result from long-term absorption of specific ultraviolet wavelengths by endogenous photosensitizers. This will be reviewed in detail below. Similar mechanisms appear to be responsible for photic retinopathy (these will also be discussed). We will consider the effect of exogenous photosensitizers on ocular tissues. Most of these photosensitizers are introduced into the circulation in pharmacologic doses for the treatment of systemic conditions often unrelated to eye conditions. Individuals may also be exposed to photosensitizers in industrial situations. Whatever the cause, the addition of exogenous photosensitizers may well change the susceptibility of ocular tissues to damage if the chemical crosses the blood–ocular barriers and if the tissues are simultaneously irradiated with wavelengths characteristically absorbed by the sensitizer.

Ocular Effects

Lid

Excessive exposure of the lids to UVB (290–315 nm) may result in acute solar erythema or "sunburn" of the eyelids. This occurs infrequently as an isolated finding because of the shading afforded by the brow and because of the normal eyelid mobility and retraction. Still, one can fall asleep while sunbathing and sustain a lid burn that may be accompanied by significant lid edema.[6] Lid edema may also be seen to accompany acute photokeratitis. (The reader is referred to the article by Forbes in this issue for extensive discussion of the cutaneous reaction to radiation.)

Conjunctiva

The thin membrane that loosely covers the sclera and lines the inner surface of the lids is called the conjunctiva. In persons who spend a lot of time out of doors, especially in solar climates, this protective membrane is exposed to considerable environmental UV. This exposure causes a thickening and hypervascularity of the conjunctiva that is analogous to solar dermal hyperkeratosis.[7] The condition is called pinguecula and can be exacerbated by other forms of physical and/or chemical irritation, such as exposure to smoke, chlorine, etc. Aside from the cosmetic problem pinguecula presents, it can lead to significant ocular discomfort and a higher susceptibility to ocular infection. This discomfort and potential for infection can be due to corneal drying from altered tear composition, and possibly from a decrease in the wiping action of the lid blink, since the elevated pinguecula effectively holds the lid off of the peripheral corneal surface (limbus).[8]

Aggravated pingueculae can develop into pterygia,[9] which are conjunctival flaps that grow onto the corneal surface and slowly migrate toward the corneal apex. Pterygia are also considered to be a direct result of UV exposure and are endemic in some equatorial climates. Both pingueculae and pterygia are usually benign and self-limiting. They benefit in their later stages from lubrication and, occasionally, surgical exicision.

Cornea

Ultraviolet effects on the cornea are perhaps the most dramatic and well-documented.[10] Patients with UV keratitis (photokeratitis, snowblindness, welder's flash) usually have a history of excessive solar exposure, especially in snowy settings or have been exposed (even briefly), without protective goggles, to the intense UV from an arc welding process. Exposures will usually be followed by a latent period of 1–24 hr. Initial symptoms include a gritty sensation and conjunctival redness progressing to pain, tearing, photophobia, and spastic lid closure. Examination shows the central corneal surface to be denuded of cells, which is the cause of the ocular pain. The condition usually resolves spontaneously within 24–48 hr, but can be aided by treatment with ocular lubricants, analgesics, topical antibiotics, and possibly patching.

Pitts[11] has reviewed and described the corneal thresholds to damage (Fig. 4). Cullen[12] has reported the effects of repetitive subthreshold doses on the cornea and described the effects ultrastructurally. The cornea absorbs most of the wavelengths below 295 nm due to the presence of nucleic acids and amino acids, which act as chromophores. Since nucleic acids have an absorption maximum[13] near 260 nm, the alterations resulting in the nuclei of epithelial cells after exposure to UVC are easily understood. Similarly, other cellular disturbances can be appreciated in view of the absorption maxima of the aromatic amino acids in the 270–288 nm range. Tryptophan absorbs also in the UVB part of the spectrum.[1]

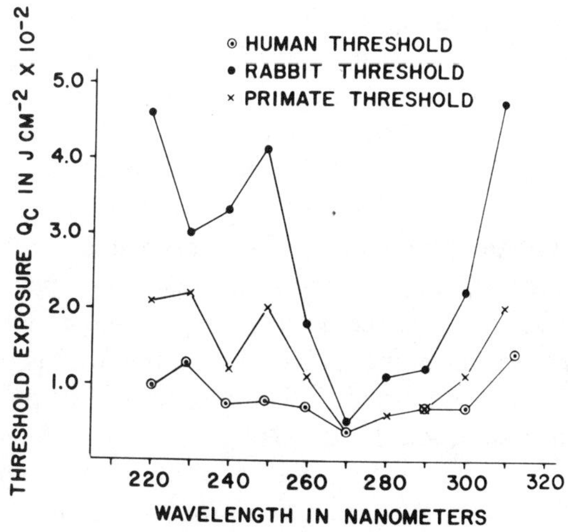

Figure 4. Comparison of the radiant exposure thresholds for the cornea of the human, the primate and the rabbit (from ref. 11).

Since the corneal epithelium is a highly regenerative tissue and there are few circumstances in which sustained damage is incurred by the patient, there has been comparatively far less basic research investigation of photodamage to this tissue than to the lens or retina.

Lens

The typical age-related changes in transparency of the lens causing increase in the scattering of light and decrease in lens transmittance have often been ascribed to lifetime exposure to sunlight. The role played by radiation in the development of the senile cataract has been extensively reviewed by Pitts.[14] While the early reports can be faulted for failing to take into account variables such as nutrition, genetics, and other socioeconomic factors in addition to radiation, Hiller,[15] in a more controlled study, reported significant association between cataract and UVB, latitude, and sunlight. Similarly, Zigman[16] described an increase in brunescent cataract with decreased latitude and an association between cataract formation and near-UV light. There is therefore some basis for investigating the role of radiation in the genesis of the senile cataract.

Some of the alterations are common with other cataractogenic processes. Loss of transparency has been described in terms of protein aggregation and cross-linking, increases in the molecular weight of protein, and loss of cellular (epithelial and fiber) organization.[17, 18] More recently, interest has been focused on the mechanisms involving the role of oxygen in the genesis of these alterations. Thus, it has become apparent that the formation of oxygen free radicals and singlet oxygen in excessive amounts (i.e., in concentrations beyond the limits of the lenticular protective mechanisms) may play a key role in the cataractogenic process, by initiating lipid peroxidation and other toxic changes.

In an excellent review of the photochemistry of proteins as this applies to the ocular lens, Grossweiner[19] has emphasized that despite reseach describing *in vitro* protein damage induced by UVC, in the *in vivo* situation we must concern ourselves with wavelengths greater than 295 nm because of the absorptive characteristics of the cornea. Thus, the initial question becomes: In the UVB and UVA range, what endogenous chromophores could initiate photochemical/photosensitized reactions? Many investigators have hypothesized that the initial intrinsic absorber is tryptophan.[19, 20] Upon exposure to UVB, photolysis occurs, resulting in the production of tryptophan free radicals and *N*-formylkynurenine (NFK). Formation of this product by sunlight was characterized by Pirie[21] and others.[22] The presence of tryptophan free radicals after UV irradiation in human lenses has also been demonstrated by Wieter[23] and Zigman.[16] Other evidence supporting the photolysis of tryptophan has been presented by Dillon;[24, 25] in these studies, the process was increased three to four times in the presence of oxygen.

Two possible mechanisms for the generation of NFK, one photolytic and one nonphotolytic, have been proposed by Grossweiner.[19] By either mechanism, the generation of NFK produces another photosensitizer with absorption characteristics in UVA (maximum 320 nm, fluorescence peak at 430 nm). According to Pileni,[26, 27] the subsequent photosensitization paths could then follow any of three ways leading to the formation of singlet oxygen, the formation of the superoxide and hydroxyl radicals, or interactions with nucleic acids and amino acids.

The resultant presence of singlet oxygen and free radicals sets the stage for numerous oxidative and nonoxidative reactions including lipid peroxidation, protein cross-linking, and enzyme inactivation. The probable role of singlet oxygen in the cataractogenic process has been addressed by Zigler and Goosey.[28] Using *in vitro* techniques, Varma and colleagues[29] have studied the role of superoxide and other oxidants. They have postulated that the generation of superoxide could occur via kinurenine or riboflavin derivatives. They[29-31] and others[32-34] have further established the presence of superoxide dismutase in lens tissue and found the protein in the lens epithelium more so than in the cortex or the nucleus. According to the authors, the relative lack of this enzyme in the latter two tissues has the net effect of extending the lifetime of superoxide and thus rendering the tissues more vulnerable. Of course other enzymes and antioxidants are theoretically available to protect the lens. Varma[29] has suggested that ascorbate may play a protective role since it is more concentrated in the aqueous humor and lens of diurnal primates than in other animals. The role of glutathione peroxidase and reductase has also been studied by Orloff[34] and Megaw.[35]

Bhuyan and Bhuyan[36] proposed a similar mechanism whereby cataractogenesis would result from the production of oxygen-reactive species and subsequent lipid peroxidation of polyunsaturated fatty acids (PUFA) (see Fig. 5). Their hypothesis extends not only to the development of the brunescent cataract but also to that of certain experimental cataracts.

In summary, many recent studies have attributed a role to the interaction of radiation and exogenous chromophores/sensitizers in the cataractogenic process. The photosensitization process eventually involves protein denaturation and lipid peroxidation. While this is an attractive hypothesis, much of the work has been done *in vitro*, and many *in vivo* studies remain to be performed to verify its legitimacy.

Retina

Despite the obvious and beneficial need for the absorption of light for stimulation of the visual pigments, Noell[37] demonstrated in 1966 that continuous exposure of the albino rat retina to light could result in damage to and destruction of this tissue. Since then many studies have addressed the nature of the damage.[38] While research into lens effects has followed biochemical and photochemical lines, up to very recently the study of retinal effects has been descriptive, using histologic and ultrastructural techniques. Clinically, it now appears that similar effects occur but that substantial loss must be incurred before the effects manifest themselves as a clinically detectable loss of visual function. Thus, it had been hypothesized that macular degeneration is the result of lifetime exposure to certain wavelengths, resulting in receptor and RPE dysfunction.

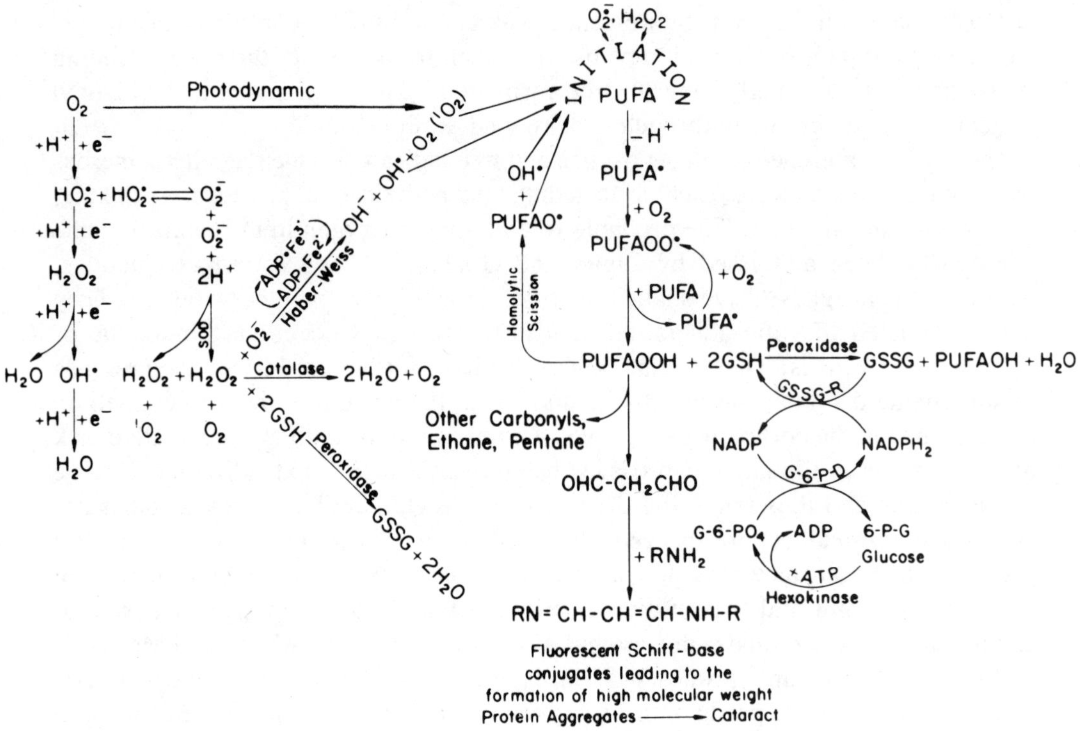

Figure 5. Mechanism of cataractogenesis proposed by Bhuyan and Bhuyan (from Molecular mechanisms of cataractogenesis II: Evidence of lipid peroxidation and membrane damage, In *Oxy-radicals and Their Scavenger Systems: Cellular and Medical Aspects*, R. Greenwald and G. Cohen, Eds., Elsevier Science Publishing Co., New York, 1982).

The pioneering work of Ham and his colleagues has differentiated two types of photochemical lesions resulting from exposure to light: one primarily related to the absorption of 400–500 nm wavelengths, the so-called blue-light lesion, which entails a primary lesion at the level of the retinal pigment epithelium; the second involves damage to the photoreceptors themselves through absorption of UV and short-wavelength light. The latter produces alterations in the rod outer segments (ROS), rod inner segments (RIS), and outer nuclear layer (ONL) with eventual loss of the cells themselves.

The blue-light damage was originally characterized clinically and histologically by Ham and colleagues.[39] According to their work, the lesion is restricted to small portions of the macula and paramacular retinal pigment epithelium and resembles the condition called solar retinitis, which was reported in military personnel during World War II. The action spectrum parallels the absorption spectrum of melanin and is inversely proportional to the wavelength. Threshold to damage was established as 30 J/cm^2 of 441-nm radiation. Immediately after exposure, little damage was observable but it became clinically visible 2 days after exposure when areas of depigmentation were noted. Histologically, this hypopigmentation can be explained by the presence in the subretinal space of numerous macrophages loaded with melanin. Their presence increased at 5–6 days. Concurrently, the cells of the RPE are swollen and some of the photoreceptors are damaged. The anomalies are gradually reduced at 10–11 days and the tissue appears close to normal at 30 days except for some depigmentation and the presence of a few macrophages. The lesion was barely distinguishable clinically 60–90 days after exposure.

The experimental conditions causing the blue-light lesion did not involve a substantial elevation in temperature in the exposed tissues. It should be emphasized that there can be thermal enhancement of photochemical lesions,[5, 40, 41] a finding which is of some concern when one is using diagnostic and surgical instruments since many sources emit both in the lower wavelengths and in the IR range. A number of filtering devices have already been designed to reduce this risk.

In a separate series of experiments, Ham and colleagues[42] have demonstrated that a threefold reduction in the radiant exposure is needed to reach threshold for this lesion when oxygen tension is increased to 217 mmHg. They have also reported preliminary data suggesting that β carotene could play a protective role in the formation of this lesion.

Crockett and Lawwill[43] have also studied the blue-light lesion, and have suggested and presented evidence for the absorption of the short-wavelength visible light (435 nm) occurring via a subcellular endogenous photosensitizer. They propose that this photosensitizer is located in the mitochondria, where numerous candidates are known to be present. The authors report that there is an oxygen dependency and have called the reaction photodynamic. (In the strict sense, photodynamic is restricted to Type II reactions involving primarily the generation of singlet oxygen and, much less efficiently, the superoxide radical. The production of superoxide as well as hydrogen peroxide and the hydroxy radical can occur via Type I reactions). The studies were conducted *in vitro* using cultures of bovine epithelium.

In many circumstances, descriptions of photic retinopathic changes have included descriptions of alterations to the RPE essentially similar to those described by Ham. However, many investigators have also described simultaneous damage to the photoreceptors.[44, 45] It is of course possible that two separate insults occur under some experimental conditions and not others, although these conditions remain to be elucidated.

Another perplexing phenomenon is that identified by Harwerth[46] and by Sperling,[47, 48] in which selective loss of sensitivity to blue stimuli following repeated intermittent exposure to intense blue light was related to a loss of blue-sensitive cones. However, irradiation with a steady blue light for 120 continuous min resulted in extensive damage to the RPE with little damage to the cones. Much more research needs to be done on the blue-light effect before it can be fully characterized and understood.

The photoreceptor damage has been coined the Noell syndrome, though this is perhaps confusing since Noell did identify both RPE and photoreceptor damage manifested at the same time under certain experimental conditions.[45] The damage consists mainly of alterations to the ROS, swelling and disruptions in the RIS, and pyknosis and loss of nuclei with sparing of the remainder of the retina in many cases. The retinal damage has been reported by a number of investigators.[40, 41, 49, 50] Some areas of the retina are more susceptible to photic insult than others, and this susceptibility is species-specific.[51] Noell has also suggested that the susceptibility to damage is age-related.[45] A third characteristic of photic insult is the role played by pigment: pigmented animals suffer less damage than do comparably treated albino animals.[52]

It is not surprising that intensity of the radiation as well as the duration of the exposure play a role in the type of lesion produced. The effect is complex, as illustrated by observations from four different experiments. Sykes and colleagues demonstrated that cyclic exposure to low levels of fluorescent light in the albino rat resulted in the death of many photoreceptors over a 30-day period.[53] Noell[45] described a situation in which illumination of the albino rat retina by 50–200 ft-c for 8–48 hr produced a loss of rod cells as well as damage to the RPE. This damage was accentuated if the test animals were maintained in constant dark for 2 weeks prior to irradiation. Conversely, if the test animals were maintained in weak cyclic light prior to irradiation, less damage was observed. The second kind of damage identified by Noell[45] involves widespread photoreceptor loss without RPE damage in young animals as a result of exposure to continuous fluorescent light for 8–50 days. This is similar to results from the studies of O'Steen and co-workers.[54, 55] Finally, Williams and colleagues[56] have recently reported no significant alteration to susceptibility in pigmented rats reared in either 3 or 100 lux, if exposed later to cyclic illumination. The role of lighting conditions to which experimental animals are exposed immediately prior to exposure, as well as that of cyclic vs. continuous exposure, has been discussed,[57, 58] but still awaits further elucidation.

Williams and colleagues[56] have also studied quantitatively the effect of a single mydriatic treatment on albino and pigmented rats when these animals were subsequently exposed to low levels of cyclic light for periods of days. Animals treated with atropine suffered worse damage than those not treated. The degree of damage was correlated to the intensity of the illumination. However, the albino retinas were comparatively more damaged than the pigmented even when the latter were dilated. For all, the superior retina was more susceptible than other parts of the retina.

The radiation effects of ultraviolet light alone on the retina are limited in most humans due to the absorptive characteristics of the cornea and the lens. However, there are situations, such as in the aphakic or pseudophakic eye, where UV absorption by the lens does not occur. Similarly, in a number of animal models the lenticular transmittance characteristics are different and permit UVA to reach the retina.

The effect of UVA on the retina has been studied by Schmidt and Zuchlich[59] and Ham and colleagues.[60] The action spectrum for retinal damage from wavelengths from 325 to 441 nm has been reported; much lower radiant exposures are needed at 325 nm to induce damage, while gradually increasing radiant exposures were needed as the wavelength increased.[60] The site of damage is primarily the photoreceptors, which undergo alterations in the outer segments, inner segments, and nuclear layer. Loss of photoreceptors occurs. Similar information on UVA-induced damage has been reported in rodents.[61]

The investigations on retinal damage have recently progressed along photochemical or biochemical lines. Delmelle[62, 63] originally proposed that retinal damage was, at least in part, the result of the activities of singlet oxygen. (The role of oxygen in the retinopathy of the premature has been well documented.[64] Since the effects of this condition are far more acute and detrimental, they are considered here as distinct from those presently under discussion.) Lion and colleagues[65] detected the formation of singlet oxygen upon irradiation of the retina. The presence of singlet oxygen suggests peroxidative effects. These effects have been studied by Anderson and colleagues.[66] Recently, Anderson and colleagues[67] extended the work of Kagan and colleagues[68] and reported on *in vitro* techniques by which they observed a 40–50% deduction in the regenerability of rhodopsin in ROS where lipid peroxidation had occurred.

The role of enzymes and antioxidants is being investigated concurrently. Joel[69] reported a decrease in vitamin E in whole retinas as a result of illumination. Similar experiments on dismutase, catalase, peroxidase, as well as other antioxidants are underway in several laboratories.[70–72]

Thus far, findings suggest that the degeneration of the ROS may result from the production of singlet oxygen and other reactive species. While further validation is needed, this hypothesis nonetheless presents a plausible schema for at least part of the damage. In contrast, the mechanisms for damage of the inner segments and nuclei have not been postulated or studied.

The relationship of photoreceptor loss to changes in acuity and other functional parameters has not yet been conclusively defined. In many situations in which damage was incurred, the visual function and psychophysical thresholds appear either to show no alteration[73–75] or can recover.[76, 77] Still other studies show a relationship between visual function tested and the degree of photoreceptor damage.[78] Henton and Sykes[79] have studied the recovery of absolute threshold with UVA-induced retinal damage in the albino rat. They compared the changes in the absolute threshold in one eye with the ultrastructural evaluation conducted on the other eye, which had been enucleated. They found that while there was profound alteration in absolute threshold following exposure, with a progressive recovery by day 7, the anatomical study showed typical photoreceptor damage immediately following exposure (up to 3 days after exposure) followed by a progressive loss in total photoreceptor population. In trying to explain these results, one could speculate that few functional changes would be observed clinically until the photoreceptor loss is extensive enough to eliminate the redundancy in the system.

Another point to consider is the level at which disruption of the rod outer segment involves incapacitation, to the point where the cell may not survive. The ability for the outer segment to be renewed implies a certain degree of protection. It has been implied that photoreceptor turnover, as well as shedding, is altered by photic insult.

Exogenous Photosensitizers

The photoeffects described so far have involved endogenous chromophores. However, a number of exogenous photosensitizers can be introduced into ocular tissues via the circulation. These are usually administered to patients in pharmacologic doses for treatment of systemic diseases unrelated to the eye.

While the number of potential photosensitizers is high because of the large number of chemicals with ring structures, only a few have been studied systematically. Of these, the best understood is 8-methoxypsoralen (8-MOP); information is presently being gathered on some of the other psoralens.

In the treatment of psoriasis a patient takes the *p*soralen drug, and after 1 or 2 hours is irradiated with a bank of *UVA* lamps.[80] Hence, *PUVA* (psoralen-UVA) is often used to describe this treatment. During irradiation, the eyes are usually protected by UV-filtering goggles and by filtering eyeglasses immediately after. The drug is removed from the circulation within 24 hr of administration. (The reader is referred to Berger's article in this issue for further information.)

The ocular effects of this treatment have been studied mainly in animal models, in which cataract formation has been documented.[81-86] Two types of alterations to lenticular tissue have been reported. Lerman[87] has reported that phosphorescence spectra from the lenses of patients treated with PUVA are identical to those of the protein photoproducts found in rats treated with PUVA. The interaction of 8-MOP with DNA and the initiation of unscheduled DNA repair in lens epithelium has been demonstrated by Jose and Yielding.[88, 89]

While the intercalation of 8-MOP with DNA and the formation of DNA cross-links is believed to be the predominating mechanism of photosensitivity,[90] this does not preclude the production of other photoproducts. Furthermore, while the DNA cross-linking mechanism is generated by Type I reactions, other 8-MOP phenomena are known to involve Type II reactions.

At the level of the retina, photoreceptor damage has been reported in both rodents and aphakic monkeys.[91-93] The type of damage is not qualitatively different from that reported for UVA alone. Thus only quantitative assessment of photoreceptor necrosis can accurately address the question of PUVA-enhanced retinal photoreceptor damage and loss.

A second class of drugs presently under investigation is the phenothiazides, which have been known for some time to exhibit photosensitizing properties. These have been investigated *in vitro* by a number of researchers, most notably Kochevar[94-96] and Testylier and colleagues.[97, 98]

It has also been known for some time that this group of drugs results in several ocular side effects.[99, 100] The task of relating any ocular effect to a particular photosensitization mechanism has been difficult. For example, pigment deposits have been noted in the cornea of patients undergoing antipsychotic therapy with these drugs. Pigment changes have also been documented in the retina, especially with the use of NP-207 and thioridazine, and, to a much lesser degree, with chlorpromazine.

Tricyclics in general, and these drugs in particular, have an affinity for melanin.[99-101] The degree to which pigment alterations, photosensitization, and visual dysfunction are associated, especially in the case of NP-207 and thioridazine, is unresolved.

However, photoproduct formation in the cornea after administration of chlorpromazine and trifluoperazine has been documented.[102, 103] Similarly, Merville[104] and Roberts[105] have identified oxygen-mediated light-dependent protein alterations in lenticular tissue, while Jose has reported the photoaddition of chlorpromazine to DNA.[89, 106] One or more of these reactions may contribute to the dose-related lenticular effects commonly reported as the "stellate-like" chlorpromazine cataract.

In the long list of potential ocular photosensitizers, the following should be noted particularly since there is biochemical *in vitro* and/or *in vivo* documentation of their photosensitizing properties: many of the quinine antimalarials; chloroquine; allopurinol used in the treatment of gout; some of the antibiotics, especially nalidixic acid and the tetracyclines; and many of the new nonsteroidal anti-inflammatory agents.[107] These compounds, as well as the psoralens and the phenothiazines, have recently been reviewed[108] for their photosensitizing properties as well as their ocular effects.

One group that must be singled out is the porphyrin deviatives currently used in cancer phototherapy. Since these molecules concentrate in malignant tissue, and have photosensitizing properties, it is easy to appreciate that they are valuable in the treatment of several cancers. Specifically because these compounds absorb in the visible part of the spectrum (which characteristic makes them uniquely powerful photosensitizers), ocular tissues must be adequately protected from unwarranted exposure to most of the optical spectrum while these compounds are systemically distributed.

The difficulty in establishing a relationship between drug effect and photosensitization is compounded by the fact that toxic reactions, independent of the photosensitizing one, must be ruled out or identified. Only recently have systematic investigations of ocular effects begun. The type of research needed to verify ocular photoeffects of drugs has been identified in the recommendations of a recent FDA workshop.[108] The information gained by the research suggested would be invaluable in assessing these effects and would certainly aid in the prevention of unwarranted ocular side effects.

Clinical Surveillance of Vision and the Eye

Persons who are at risk for ocular damage or functional visual loss from light exposure, drug toxicity, or drug–light interaction should be considered for ocular and visual surveillance testing.[1, 99, 109, 110] Many clinical and research techniques are available,[111-113] although their use in assessing photic and photosensitized visual effects has thus far been limited. Much of the reported work in surveillance has been to detect ocular disease and toxicity, which may have effects similar to photodamage. Therefore, established ocular and visual testing regimens will be discussed here in relation to their ability to detect early functional or structural eye changes, whether such changes are caused by light, drugs, or their combined effects.

In individuals exposed to excessive UV and short visible wavelengths or in those taking toxic or phototoxic substances, a thorough ocular evaluation is valuable, as are periodic follow-up examinations to detect any changes. When this surveillance is rigorously pursued, using the most appropriate techniques, it is possible to discover even subtle alterations early enough to change or cease treatment and/or exposure, thus reversing the deleterious effects.[99, 114]

While this approach seems logical, the practical ocular and visual surveillance of at-risk individuals is frequently complicated by a number of problems. For example, frequently clinicians have no knowledge of a patient's work environment. Similarly, many physicians who prescribe a given drug are often unaware of the visual disturbances associated with its use or for the need for follow-up eye evaluation. When such evaluation is sought, it is frequently conducted at a very basic level, using routine clinical procedures that may not detect visual disturbances until their later stages, when they are frequently irreversible.[115–119] The proper surveillance of a patient must employ the clinical,[111] and, when necessary, research procedures[120] most appropriate for detecting both known and suspected drug and light effects in their early stages.

The following sections discuss the ocular and visual surveillance tools currently available to the ophthalmic practitioner. The general use of each and its relationship to known toxic and disease effects will be detailed.

Importance of the Case History

The findings taken in a carefully documented case history are often among the most important in a patient's record. A thorough medical interview can reveal a work environment or the presence of physical and ocular conditions in the patient that would automatically place him or her at higher risk.[121–124] Any early sign of cataracts, macular drusen, or macular degeneration also identify the patients who must be carefully evaluated over time, since these conditions may enhance photic absorption or indicate compromised cellular physiology.[1] Knowledge of the patient's medications is very important as a source of toxic effects,[99] photosensitized responses,[81–83, 138] and drug interactive effects. Certain drugs, including chloroquine, are known to have toxic effect; and specific clinical tests have been recommended during their administration to monitor ocular and visual changes.[99, 125–137] Finally, since the Framingham study has pointed out a relationship between eye color and macular degeneration, it would be useful prospectively to note eye color and skin pigmentation.

Visual Resolution

Standard Snellen letter visual acuity is traditionally used to measure the central visual resolution capability. Gross disruption of the central retina (macular hole, macular scar, hemorrhage) or of the lens (cataract) is usually required to produce a measurable reduction in acuity.[139, 140] This makes the Snellen test relatively poor for the detection of the early or more subtle effects in toxicity or disease.

An example of this problem in detecting ocular conditions is macular edema, which is seen during the early stages of macular degeneration[141] and in phototoxicity.[142] Patients with early macular edema will frequently experience no loss in visual acuity.[140] The sensitivity of acuity determination can, however, be enhanced by the addition of the "photostress test."[143, 144] In this procedure, a bright light is used to create an after-image that temporarily reduces the visual acuity. The recovery time from this test is extended when macular edema is present.

Measurement of contrast sensitivity function (CSF) is another recently developed resolution test that addresses the problem of early detection.[140, 145–149] This method has been shown to be more subtle than Snellen resolution for documenting functional visual loss due to eye disease, especially in the earlier stages.[148–151] In addition to detecting resolution losses earlier, CSF provides more information to the examiner by allowing the evaluation of a broader range of image dimensions. This is important because the vision system is thought to be mediated by more than one spatial "channel" for image resolution.

These channels have been described psychophysically[152] and electrophysiologically[153] and are related to the size of the target being observed. Spatial frequency is used to describe the size of such targets. High spatial frequency targets are small or have small patterns of repetition, such as a fine sine wave grating (Fig. 6A). Low spatial frequency targets are generally larger or have larger repeat patterns (Fig. 6B).

These two types of targets are received by the visual system with differing levels of sensitivity so that the amount of contrast required for their detection varies across the spectrum of spatial frequencies to create what is called CSF (Fig. 7). Many predictable alterations in this function are found in certain eye diseases[151, 152, 155–157] that may share causative features with toxic and photic phenomena and therefore may also be useful for surveillance. Because of its relative newness in eye care and vision research, CSF has not been used extensively in the study of photic or toxic effects,

Perimetry

Perimetry is an effective tool for documenting changes in the extent and relative sensitivity of the visual field.[158] Such changes can be very important in the evaluation of patients experiencing neurologic changes that restrict the visual field or depress the overall retinal threshold. Perimetry is excellent for detecting early signs of toxicity from chloroquine[99, 127, 159] and NP-207 therapy.[160] Chloroquine creates paracentral scotomas that are reversible in the early stages, but which lead to macular degeneration unless the drug is discontinued.[123, 159]

Unfortunately, although perimetry is available in most clinics, it is often not used in routine clinical practice. This may change once recently developed computerized field testing devices become more widely available.[161]

In addition to the use of classic white-light perimetry, which uses relatively dim targets and backgrounds, there has been interest in the past decade in the measurement

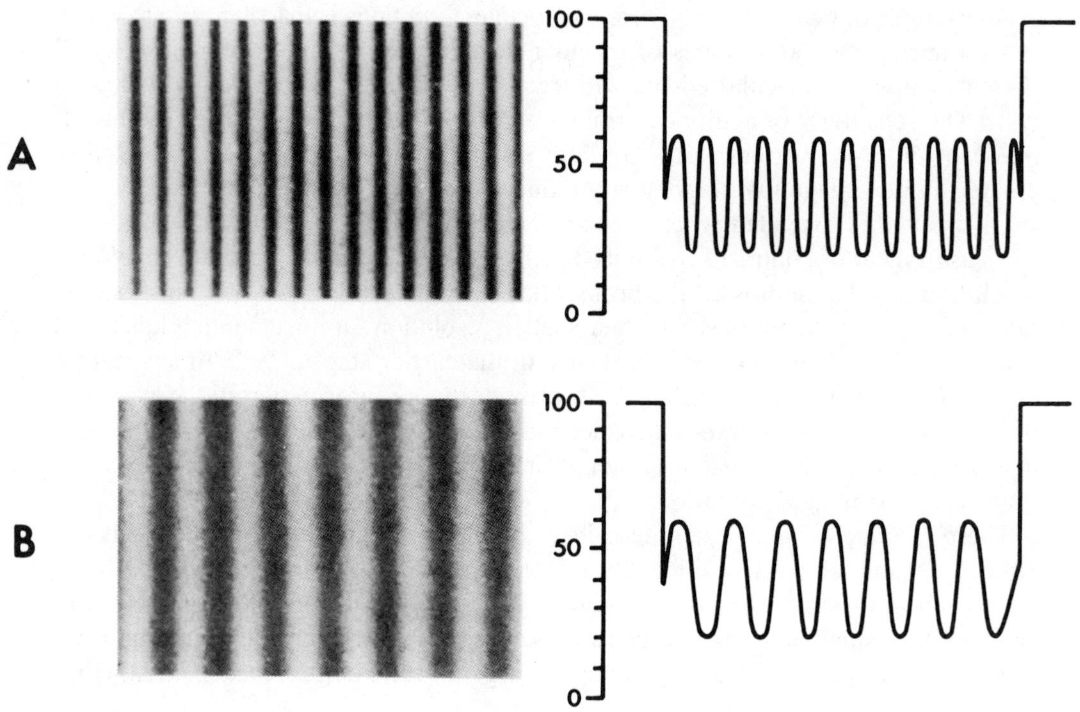

Figure 6. Examples of high (A) and low (B) spatial frequency sine wave spatial gratings (from ref. 148).

of thresholds of high-intensity spectral (color) increment detection. These measurements have been taken in the macular and in the peripheral retina as a functional determination of the cone system (red, green, blue). There appears to be selective depression of one or more systems in certain types of retinal and/or neural pathology, which indicates the retinal area or the visual pathway component being affected.[113, 162–169]

Dark Adaptometry

Dark adaptometry[170–173] is a technique that has been used to describe the development and progression of diseases causing night blindness. The procedure is predicated upon the fact that in the normal retina the maximum sensitivity to light is attained after approximately 30 min through a process called dark adaptation. The maximum sensitivity attainable is reduced in cases where the rod function is depressed. Examples of this type of toxicity are found with the use of chloroquine,[127, 174, 175] thioridazine,[176] alcohol,[177] and other agents.[178]

	Age range (years)	Mean age (years)	Number of eyes tested	Mean Snellen acuity	Mean pupil diam (mm)
(- - -)	8-15	13	3	0.8	6.3
(— —)	18-39	23	25	1.2	5.4
(· · ·)	45-66	58	5	0.7	3.8

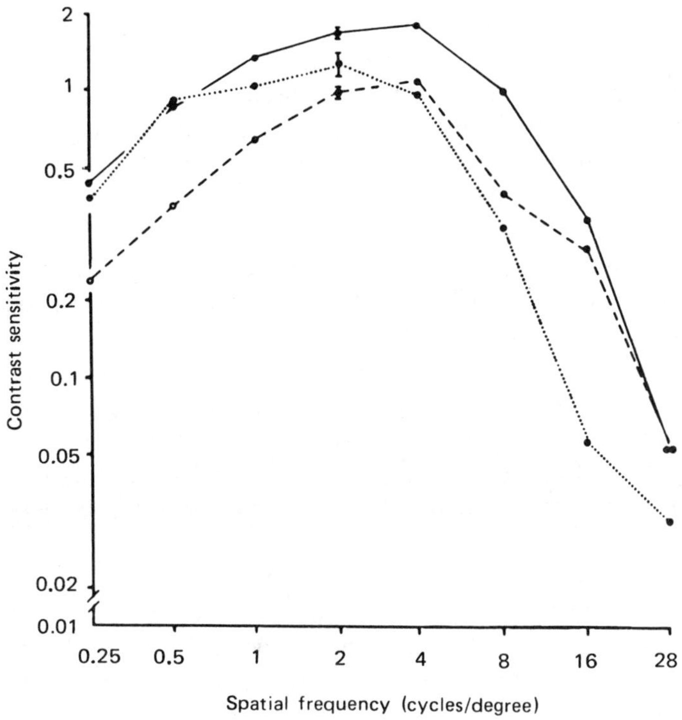

Figure 7. Examples of contrast sensitivity functions (CSF) for observers of different ages (from K. Arundale, An investigation into the variation of human contrast sensitivity with age and ocular pathology, *Brit. J. Ophthalmol. 62*:213, 1978).

Color Vision Testing

The use of color vision testing is also significant in the monitoring of toxic and photic eye disease. Much has been written about the characteristic changes in color vision,[179-185] which are a sensitive indicator in functional retinal change. Unfortunately, although clinical color test plates (pseudoisochromatic plates), are quick and easy to use, they are not sensitive to subtle color anomalies and are rarely designed to test for blue–yellow defects, which are most commonly expected to occur in retinal toxicity.[179] Furthermore, plate tests do not quantify the level of a color disturbance, which is essential for monitoring progress in surveillance testing.

In contrast, more specialized though not frequently used, color matching (Farnsworth-Munsell, 100 Hue Test[186]), and color mixing procedures (Nagel[179, 182] and Pickford-Nicolson[187, 188] Anomaloscopes), are sensitive to any type of color deficiency axis and can be used to quantify even subtle defects. Since a retinal alteration may selectively affect one or more of the cone mechanisms, it is logical to test color function comprehensively in patients who are at risk.

Electrodiagnostic Testing

Just as color vision testing, perimetry, and dark adaptation allow assessment of the various retinal components within the visual field (i.e., rods; and red, green, and blue cones), electrodiagnostic testing can be used similarly to evaluate the vision system and to determine the level of the visual system (from photoreceptor to visual cortex) at which functional visual loss occurs.[111, 189] Electrophysiological testing is also highly important to research (as was demonstrated by Noell, among others[45]) because the ERG and the VER are objective tests and can therefore be used in animal models to assess visual function indirectly, without having to resort to animal psychophysics.

The electroretinogram (ERG) is a flash- or pattern-simulated electrical response of the retina–RPE system.[190–192] Because the ERG has a finite time course, it is possible to estimate the level of the retinal–neural pathway at which a toxic effect has occurred.[193–196] The RPE seems to be the retinal tissue that is mainly affected in chloroquine retinopathy, as judged by a disappearance of the c-wave and a decrease in the b-wave amplitude of the ERG.[130, 136, 174, 197] The ERG is relatively easy to collect and requires little patient cooperation. It is possible to quantify the measurement better through computer averaging of multiple traces. One drawback to the technique is that it measures the mass response of the retina to the light stimulus. This fact may prevent detection of relatively small retinal damage areas, such as the macula. For this reason, the technique has been used foveally[198, 199] and with a patterned stimulus to enhance its resolution in the measurement of macular function.[200–203]

A measure of the standing potential of the eye, the electrooculogram (EOG) also serves as an indicator of the RPE integrity.[204, 205] Toxic effects[206] in the RPE, such as occur with the use of chloroquine[207–209] and indomethacin[210] will alter the amplitude of this determination. The EOG is, however, more difficult to use than the ERG, and is subject to a great deal of variability due to differences in technique, as well as the need for patient cooperation. Because of these reasons, several authors have questioned the value of this technique for early detection of chloroquine retinopathy.[130, 136, 137, 211]

The visually evoked response (VER)[212, 213] is an electroencephalographic recording over the visual area of the brain in response to a flash or patterned stimulus. Since the macula is highly amplified in its representation within the cortex, a relatively significant electrical response corresponding to the central area of the retina can be measured by this technique. For this reason, the VER can be more effective than the ERG for objectively determining the effects of macular disease. This technique also can be used to determine latency of arrival of the signal to the cortex, which can indicate toxicity to the neural transmission pathway.[214] These factors make this test an excellent choice for monitoring patients at risk for macular and/or optic nerve diseases.

Ophthalmoscopy/Biomicroscopy

Ophthalmoscopy is the procedure by which the examiner can look into the ocular fundus to detect abnormalities in structure that may be visible clinically. Ophthalmoscopy can be performed by either of two means, referred to as the direct and indirect techniques, each with its own advantages for certain purposes.[215] To be visible ophthalmoscopically, an abnormality must be large enough that it can be resolved using these relatively low-magnification *in vivo* techniques. Of course, this limitation allows the observation of only gross effects and prevents evaluation of damage at a cellular or subcellular level.[216] The examiner's ability to monitor the appearance of the fundus from one visit to the next may also be limited due to the problems of memory and imprecision of drawings.

An ophthalmic biomicroscope is another technique used to evaluate the eye for physical changes. Although designed primarily for external ocular evaluation, the biomicroscope can be effectively used to study the fundus as well. The use of a Hruby lens, Goldman three-mirror lens, or a high powered (60–90 D) condensing lens enables the examiner to obtain a very clear view of the posterior vitreous and retina. This view can surpass that obtained with the opthalmoscope, but it is still of a relatively low magnification, and significant retinal alteration must occur in order to be detected. The same limitations of accuracy of recording as occur with the ophthalmoscope are seen with this instrument.

Ocular Photographic Techniques

In contrast to ophthalmoscopy and biomicroscopy, internal and external photography can be effectively used to document accurately the gross appearance of a lesion large enough to see be visualized *in vivo*. It does not, however, increase the resolution ability afforded the examiner compared to other fundus exam techniques. In fact, routine photographic methods typically are reduced in their resolution capabilities due to the inherent limitations of the film grain. This problem is currently being addressed by application of recently developed computerized image-enhancement techniques. The applications of this technology are only in their beginning stages and have not been utilized in toxicology as yet.

Photography can be enhanced by the use of intravenous injection of fluorescent dyes.[141, 217] These dyes accent the flow characteristics of blood within the vessels and demonstrate fluid leakage, drusen formation, and other abnormalities in the surrounding retina that may indicate phototoxic retinal changes.[142] One problem with angiography is that it is an invasive procedure with a certain amount of associated morbidity.[218] Therefore, its clinical use is usually reserved for patients already exhibiting other signs and symptoms.

Quantitative estimates of intraocular fluorescence can be made using a radiometric device mounted within the eyepiece of a biomicroscope or fundus camera.[219-221] This is useful to quantify the amount of dye leakage that may have occurred during an angiográm or to detect another fluorescent compound that may be of interest. It may also be used to measure the amount of naturally occurring fluorescent materials[222] within the eye (especially the lens) that occur with aging or physiological decompensation.

A similar technique, called Scheimpflug photography,[223, 224] is used to quantify *in vivo* changes in the lens. This procedure combines the use of radiometric measurement and computerized image processing to enable the examiner to create an anterior-to-posterior profile of lens fluorescence and/or transmittance *in vivo.*

Conclusion

As discussed throughout this article, certain individuals are at higher risk of ocular photodamage because of aphakia, exposure to higher-energy illuminations, or due to the presence of environmental or medicinal photosensitizers. For these persons, it is important to consider the use of protective lenses[225] to block the passage of certain wavelengths. Ongoing discussions among researchers in the field point to the need to generally restrict the access of UV and in some cases short-wave visible energy to the eye.[5] Certainly, patients unergoing phototherapy such as PUVA treatment need such protection.[226] There is no doubt that continued research in this area will significantly improve the availability and use of such protective measures to the benefit of those at risk.

References

1. S. Lerman, *Radiant Energy and the Eye*, Macmillan, New York, 1980.
2. A. A. Lamola and N. J. Turro, Spectroscopy, In *The Science of Photobiology*, K. C. Smith, Ed., Plenum Rosetta, New York, 1977.
3. J. D. Spikes, Photosensitization in mammalian cells, In *Photoimmunology*, J. A. Parrish, M. L. Kripke, and W. L. Morison, Eds., Plenum Press, New York, 1983.
4. C. V Parke, Mechanisms of chemical toxicity—A unifying hypothesis, *Reg. Tox. Pharm. 2*:267, 1982.
5. W. T. Ham, Jr., R. G Allen, L. Feeney-Burns, M. F. Marmor, L. M. Parver, P. H. Proctor, D. H. Sliney, and M. L. Wolbarsht, Involvement of the retinal pigment epithelium, In *Optical Radiation and Visual Health*, M. Waxler and V. M. Hitchins, Eds., CRC Press, Boca Raton.
6. S. E. Bloomfield, Clinical allergy and immunology of the external eye, In *Clinical Ophthalmology*, Vol. 5, T. D. Duane, Ed., Harper & Row, Philadelphia, 1984.
7. G. K. Klintworth, Chronic active keratopathy: A condition associated with conjunctional elastosis (pingueculae) and typified by characteristic extracellular concretions, *Amer. J. Pathol. 67*:327, 1972.

8. F. T. Fraunfelder, C. Hanna, and J. M. Parker, Spheroid degeneration of the cornea and conjunctiva: I. Clinical course and characteristics, *Amer. J. Ophthalmol. 74*:821, 1972.
9. R. G. Freeman and J. M. Know, Ultraviolet-induced corneal tumors in different species and strains of animal, *J. Invest. Derm. 43*:431, 1964.
10. W. Buschke, J. S. Friedenwal, and S. G. Moses, Effects of UV irradiation on corneal epithelium: Mitosis, nuclear fragmentation, post-traumatic cell movement, loss of tissue cohesion, *J. Cell. Comp. Physiol. 32*:161, 1948.
11. D. G. Pitts. A. P. Cullen, and W. H. Parr, Ocular ultraviolet effects in the rabbit eye, DHEW (NIOSH) Publication No. 77-130, 1976.
12. A. P. Cullen, Additive effects of ultraviolet radiation, *Amer. J. Optom. Physiol. Opt. 57*:808, 1980.
13. R. R. Anderson, Tissue optics in photoimmunology, In *Photoimmunology*, J. A. Parrish, M. L. Kripke, and W. L. Morison, Eds., Plenum Press, New York, 1983.
14. D. G. Pitts, L. L. Cameron, J. Jose, S. Lerman, E. Mass, S. Zigler, S. Zigman, and J. Zulich, Optical Radiation and Cataracts, In *Optical Radiation and Visual Health*, M. Waxler and V. M. Hitchins, Eds., CRC Press, Boca Raton, in press.
15. R. Hiller, R. D. Sperduto, and F. Ederer, Epidemiologic associations with cataract in the 1971–1972 National Health and Nutrition Examination Survey, *Amer. J. Epidemiol. 118*:239, 1983.
16. A. Zigman, The role of sunlight in human cataract formation, *Surv. Ophthalmol. 27*:317, 1983.
17. J. A. Jedziniak, J. H. Kinoshita, E. M. Yates, L. O. Hocker, and G. B. Benedek, On the presence and mechanisms of formation of heavy molecular weight aggregates in human and normal cataractous lenses, *Exp. Eye Res. 15*:185, 1973.
18. S. Lerman, Biophysical aspects of corneal and lenticular transparency, *Curr. Eye Res. 3*:3, 1984.
19. L. I. Grossweiner, Photochemistry of proteins: A review, *Curr. Eye Res. 3*:137, 1984.
20. D. Grover and S. Zigman, Coloration of human lenses by near UV photo-oxidized tryptophan, *Exp. Eye Res. 13*:70, 1972.
21. A. Pirie, Formation of *N*-formylkynurenine in proteins from lens and other sources of exposure to sunlight, *Biochem. J. 125*:203, 1971.
22. R. Van Heyningen, Photo-oxidation of lens proteins by sunlight in the presence of fluorescent derivatives of kynurenine isolated from the human lens, *Exp. Eye Res. 17*:137, 1973.
23. J. J. Weiter and E. D. Finch, Paramagnetic species in cataractous human lenses, *Nature 254*:536, 1975.
24. J. Dillon and A. Spector, A comparison of aerobic and anaerobic photolysis of lens protein, *Exp. Eye Res. 31*:591, 1980.
25. J. Dillon, Photolytic changes in lens proteins, *Curr. Eye Res. 3*:145, 1984.
26. M. P. Pileni, P. Walrant, and R. Santus, Electronic properties of *N*-formylkynurenine and related compounds, *J. Phys. Chem. 80*:1804, 1976.
27. M. P. Pileni, R. Santus, and E. J. Land, On the photosensitized properties of *N*-formylkynurenine and related compounds, *Photochem. Photobiol. 28*:525-529, 1978.
28. J. S. Zigler and J. D. Goosey, Singlet oxygen as a possible factor in human senile nuclear cataract development, *Curr. Eye Res. 3*:59, 1984.
29. S. D. Varma, D. Chand, Y. R. Sharma, J. F. Kuck, Jr., and R. D. Richards, Oxidative stress on lens and cataract formation: Role of light and oxygen, *Curr. Eye Res. 3*:35, 1984.
30. S. D. Varma, T. Ets, and R. D. Richards, Superoxide dismutase in the lens, *Invest. Ophthalmol. Vis. Sci. 16* (ARVO Suppl.):15, 1977.
31. S. D. Varma, T. Ets, and R. D. Richards, Protection against superoxide radical in rat lens, *Ophthalm. Res. 9*:421–431, 1977.
32. J. F. Kuck, Jr., Activity of superoxide dismutase in rat lens epithelium and in fiber, *Invest. Ophthalmol. Vis. Sci. 16* (ARVO Suppl.):15, 1977.
33. D. K. Bhuyan and K. C. Bhuyan, Superoxide dismutase in calf lens, *Invest. Ophthalmol. Vis. Sci. 16* (ARVO Suppl.):15, 1977.

34. C. Ohrloff, O. Hackwin, R. Olson, and S. Dickman, Glutathione peroxidase, Glutathione reductase and superoxide dismutase in the aging lens, *Curr. Eye Res. 3*:109, 1984.
35. J. M. Megaw, Glutathione and ocular photobiology, *Curr. Eye Res. 3*:88, 1984.
36. K. Bhuyan and D. K. Bhuyan, Molecular mechanism of cataractogenesis: III. Toxic metabolites of oxygen as initiators of lipid peroxidation and cataract, *Curr. Eye Res. 3*:67–81, 1984.
37. W. K. Noell, V. S. Walker, B. S. Kong, and S. Berman, Retinal damage by light in rats, *Invest. Ophthalmol. Vis. Sci. 5*:450, 1966.
38. J. Lanum, The damaging effects of light on the retina: Empirical findings, theoretical and practical implications, *Surv. Ophthalmol. 22*:221, 1978.
39. W. T. Ham, J. J. Ruffolo, H. A. Mueller, and D. Guerry, III, The nature of retinal radiation damage: Dependence on wavelength power level and exposure time, *Vis. Res. 20*:1105–1111, 1980.
40. E. Friedman and T. Kuwabara, The retinal pigment epithelium: The damaging effects of radiant energy, *Arch. Ophthalmol. 80*:265, 1968.
41. M. O. M. Tso, B. S. Fine, and L. E. Zimmerman, Photic maculopathy produced by the indirect ophthalmosocpe, *Amer. J. Ophthalmol. 73*:686, 1972.
42. W. T. Ham, Jr., H. A. Mueller, J. J. Ruffolo, Jr., J. E. Millen, S. F. Cleary, R. K. Guerry, and D. Guerry, III, Basic mechanisms underlying the production of photochemical lesions in the mammalian retina, *Curr. Eye Res. 3*:165, 1984.
43. R. S. Crockett and T. Lawwill, Oxygen dependence on damage by 435 nm light in cultured retinal epithelium, *Curr. Eye Res. 3*:209, 1984.
44. D. Fuller, R. Machemer, and R. W. Knighton, Retinal damage produced by intraocular fiber optic light, *Vis. Res. 20*:1055–1072, 1980.
45. W. K. Noell, Possible mechanisms of photoreceptor damage by light in mammalian eye, *Vis. Res. 20*:1163, 1980.
46. R. S. Harwerth and H. C. Sperling, Prolonged color blindness induced by intense spectral lights in Rhesus monkeys, *Science 174*:520, 1971.
47. H. C. Sperling, C. Johnson, and R. S. Harwerth, Differential spectral photic damage to primate cones, *Vis. Res. 20*:1117, 1980.
48. H. C. Sperling and C. Johnson, Histological findings in the receptor layer of the primate retina associated with light-induced dichromacy, *Mod. Probe Ophthalmol. 13*:291, 1974.
49. R. Gorn and T. Kuwabara, Retinal damage by visible light. A physiologic study, *Arch. Ophthalmol. 77*:115, 1967.
50. S. Zigman and T. Vaughn, Near-ultraviolet effects on the lenses and retinas of mice, *Invest. Ophthalmol. 13*:462, 1974.
51. T. P. Williams and B. N. Baker, *The Effects of Constant Light on Visual Processes*, Plenum Press, New York, 1980.
52. L. M. Rapp and T. P. Williams, The role of ocular pigmentation in protecting against retinal light damage, *Vis. Res. 20*:1127–1131, 1980.
53. S. M. Sykes, W. J. Robinson, Jr., and J. C. Bieri, Retinal damage by cyclic light and the effect of vitamin E, DHHS Pub. FDA, 81:8156, 1981.
54. W. K. O'Steen and K. V. Anderson, Photoreceptor degeneration after exposure of rats to incandescent illumination, *Z. Zellforsch, 127*:306–313, 1972.
55. W. K. O'Steen, Hormonal and dim light effects in retina photo-damage, *Photochem. Photobiol. 29*:745, 1973.
56. R. A. Williams, A. C. Howard, and T. P. Williams, Retinal damage in pigmented and albino rats exposed to low levels of cyclic light following a single mydriatic treatment, *Curr. Eye Res. 4*:97, 1985.
57. D. T. Organisciak and W. K. Noell, The rod outer segment phospholipid/opsin ratio of rats maintained in constant dark or cyclic light, *Invest. Ophthalmol. Vis. Sci. 16*:188, 1977.
58. W. K. Noell and R. Albrecht, Irreversible effects of visible light on the retina: Role of vitamin A, *Science 172*:76, 1971.

59. R. E. Schmit and J. A. Zuclich, Retinal lesions due to ultraviolet laser exposure, *Invest. Ophthalmol. Vis. Sci. 19*:1166, 1980.
60. W. T. Ham, H. A. Mueller, J. J. Ruffolo, D. Guerry, and R. K. Guerry, Action spectrum for retinal injury from near ultraviolet radiation in the aphakic monkey, *Amer. J. Ophthalmol. 93*:299, 1982.
61. S. Zigman and T. Vaughan, Near-ultraviolet effects on the lenses and retinas of mice, *Invest. Ophthalmol. 13*:462, 1974.
62. M. Delmelle, Retinal damage by light: Possible implication of singlet oxygen, *Biophys. Struct. Mech. 3*:195, 1977.
63. M. Delmelle, An investigation of retinal as a source of singlet oxygen, *Photochem. Photobiol. 27*:731, 1978.
64. A. Patz, Current concepts of the effect of oxygen on the developing retina, *Curr. Eye Res. 3*:159, 1984.
65. Y. Lion, M. Demelle, and A. Van de Vorst, New method of detecting singlet oxygen production, *Nature 263*:442, 1976.
66. R. E. Anderson, L. M. Rapp, and R. D. Wiegand, Lipid peroxidation and retinal degeneraton, *Curr. Eye Res. 3*:223, 1984.
67. R. E. Anderson, M. B. Maude, and J. C. Nielsen, Effect of lipid peroxidation on rhodopsin regeneration. *Curr. Eye Res. 4*:65, 1985.
68. V. Kagan, A. Shvedova, K. Novikov, and Y. Kozlov, Light-induced free radical oxidation of membrane lipids in photoreceptors of frog retina, *Biochem. Biophys. Act. 330*:76, 1973.
69. C. Joel, S. Briggs, D. Goal, J. Hannor, M. Kahlow, M. Stein, A. Tarver, and A. Yip, Light causes early loss of retinal tocopheral *in vivo. Invest Ophthalmol. Vis. Sci.* (ARVO Suppl.):116, 1981.
70. W. G. Robison, T, Kuwabara, and J. G. Bieri, The roles of vitamin E and unsaturated fatty acids in the visual process,*Retina 3*:263, 1982.
71. B. J. Tripathi and R. C. Tripathi, Cellular and subcellular events in retinopathy of oxygen toxicity with a preliminary report on the preventive role of vitamin E and gamma-aminobutyric acid: A study *in vitro, Curr. Eye Res. 3*:193, 1984.
72. W. T. Ham, Jr., Personal communication.
73. K. V. Anderson and W. K. O'Steen, Black–white discrimination in rats without photoreceptors, *Exp. Neurol. 34*:446, 1972.
74. K. V. Anderson and W. K. O'Steen, Altered response latencies on visual discrimination tasks in rats with damaged retinas, *Physiol. Behav. 12*:633, 1974.
75. M. H. Bennet, R. F. Dyer, and J. D. Dunn, Visual-deficit following long-term light exposure, *Exp. Neurol. 38*:80, 1973.
76. M. Tso, Photic maculopathy in rhesus monkey, Invest. Ophthalmol. 12:17, 1973.
77. R. Penner and J. N. McNair, Eclipse blindness, *Amer. J. Ophthalmol. 81*:1452–1457, 1966.
78. W. W. Henton and S. M. Sykes, Changes in absolute threshold with light-induced retinal damage, *Physiol. Behav. 31*:179–185, 1983.
79. W. E. Henton and S. M. Sykes, Recovery of absolute threshold with UV-A induced retinal damage, *Physiol. Behav. 32*:949, 1984.
80. J. A. Parrish, T. B. Fitzpatrick, L. Tanenbaum, and M. A. Pathak, Photochemotherapy of psoriasis with oral methoxalen and longwave ultraviolet light, *New Engl. J. Med. 291*:1207, 1974.
81. T. M. Cloud, R. Hakim, and A. C. Griffin, Photosensitization of the eye with methoxalen I. Acute effects, *Arch. Ophthalmol. 64*:346, 1960.
82. T. M. Cloud, R. Hakim, and A. C. Griffin, Photosensitization of the eye with methoxalen II: Chronic effects, *Arch. Ophthalmol. 66*:689, 1961.
83. R. G. Freeman and D. Troll, Photosensitization of the eye by 8-methoxypsoralen, *J. Invest. Derm. 53*(6):449, 1969.
84. I. Milkuni, T. Fujiwana, and H. Obazawa, The effects of photosensitizing substance on the rat lens: The effect of 8 MOP on the rat lens epthelial cells, *Nippon Ganka Gakkai Zasski 81*:1412, 1977.

85. H. C. Wulf and M. P. Andreasen, Concentration of ^{3}H-8-methoxy psoralen and its metabolites in the rat lens and eye after a single oral administration, *Invest. Ophthalmol. Vis. Sci. 22*:32, 1982.
86. J. G. Jose, H. R. Koch, and A. Respondek, Histologic observations of the lenses of psoralen and UVA treated albino rats and a theory as to underlying mechanism, *Graefes Arch. Clin. Exp. Ophthalmol. 219*:44, 1982.
87. S. Lerman, J. Megaw, and K. Gardner, PUVA: Psoralen-longwave ultraviolet therapy and human cataractogenesis, *Invest. Ophthalmol. Vis. Sci. 23*(6):801, 1982.
88. J. G. Jose and K. L. Yielding, Unscheduled DNA synthesis in lens epithelium following ultraviolet irradiation, *Exp. Eye Res. 24*:113, 1977.
89. J. G. Jose and K. L. Yielding, Photosensitive cataractogens, chlorpromazine and methoxalen cause DNA repair synthesis in lens epithelial cells, *Invest. Ophthalmol. Vis. Sci. 17*(7):687, 1978.
90. M. A. Pathak and P. C. Joshi, The nature and molecular basis of cutaneous photosensitivity reaction to psoralens and coal tar, *J. Invest. Derm. 80*:66, 1983.
91. S. Lerman, J. Negaw, K. Gardner, Y. Taker, Y. Franks, and A. Gammon, Photobinding of ^{3}H-8-methoxypsoralen to monkey intraocular tissues. *Invest. Ophthalmol. Vis. Sci. 25*:1267, 1984.
92. P. Dayhaw-Barker, F. M. Barker, II, and P. D. Forbes, Retinal effects of short term exposure to 8-MOP and UVA, *Photochem. Photobiol. 37* (Suppl.):83, 1983.
93. P. Dayhaw-Barker, F. M. Barker, II, and K. Diebert, Effects of three drugs on the retinal threshold to damage. *Amer. J. Optom. Physiol. Opt. 59*(10):16, 1982.
94. I. E. Kochevar and A. A. Lamola, Chlorpromazine and protriptyline phototoxicity: Photosensitized oxygen-independent cell hemolysis, *Photochem. Photobiol. 29*:791, 1979.
95. I. E. Kochevar, Phototoxicity mechanisms: Chlorpromazine photosensitized damage to DNA and cell membranes, *J. Invest. Derm. 77*(1):59, 1981.
96. I. E. Kochevar and J. Horn, Photoproducts of chlorpromazine which cause red blood cell lysis, *Photochem. Photobiol. 37*(2):163, 1983.
97. G. Testylier, D. Daveloose, F. Leterrier, O. Buchmann, and M. Shimoni, Photochemical binding of phenothiazines on bioligical membranes proteins, *Photochem. Photobiol. 39*(3):273, 1984.
98. G. Testylier, G. Clement, and F. Leterrier, Photochemical interactions of chlorpromazine with phospholipids and cholesterol in artificial and natural membranes, *Photochem. Photobiol. 39*(3):277, 1984.
99. W. Meier-Ruge, Drug-induced retinopathy, *CRC Crit. Rev. Toxicol. 325*: 1972.
100. A. M. Potts, The concentrations of phenothiazines in the eye of experimental animals, *Invest. Ophthalmol. 1*:522, 1962.
101. A. M. Potts, Uveal pigment and phenothiazine compounds, *Trans. Amer. Ophthalmol. Soc. 60*:517, 1962.
102. D. S. Hull, S. Csukas, and K. Green, Chlorpromazine-induced corneal endothelial phototoxicity, *Invest. Ophthalmol. Vis. Sci. 22*:502, 1982.
103. D. S. Hull, S. Csukas, and K. Green, Trifluoperazine: Corneal endothelial phototoxicity, *Photochem. Photibol. 38*(4):425, 1983.
104. M. P. Merville, J. Decuyper, J. Pietle, C. M. Calbert-Bacg, and A. Van de Vorst, *In-vitro* cross-linking of bovine lens proteins photosensitized by promazines, *Invest. Ophthalmol. Vis. Sci. 25*:573, 1984.
105. J. E. Roberts, The photodynamic effect of chlorpromazine, promazine and hematoporphyin on lens protein, *J. Invest. Ophthalmol. Vis. Sci. 25*:748, 1984.
106. J. G. Jose, Photomutagenesis by chlorinated phenothiazine tranquilizers, *Proc. Natl. Acad. Sci. 76*:469, 1979.
107. I. E. Kochevar, K. W. Hoover, and M. Gawienowski, Benoxaprofen photosensitization of cell membrane disruption, *J. Invest. Derm. 82*:214, 1984.
108. P. Dayhaw-Barker, P. D. Forbes, D. Fox, S. Lerman, J. McGinness, J. Megaw, M. Waxler, and R. Felten, Drug phototoxicity and visual health, In *Optical Radiation and Visual Health*, M. Waxler and V. M. Hutchins, Eds., CRC Press, Boca Raton, in press.

109. W. T. Grant, *Toxicology of the Eye*, Charles C. Thomas, Springfield, Ill., 1963.
110. F. T. Fraunfelder, *Drug-Induced Ocular Side Effects and Drug Interactions*, 2nd ed., Lea & Febiger, Philadelphia, 1982.
111. A. E. Krill, *Hereditary Retinal and Choroidal Diseases*, Vol. 1, Harper & Row, Hagerstown, MD, 1972.
112. E. Zrenner, Ed., Special tests of visual function, basic problems and clinical applications, In *Developmental Ophthalmology*, Vol. 9, Karger, Basel, 1984.
113. G. Verriest and A. Uvijls, Spectral increment thresholds on a white background in different age groups of normal subjects and in acquired ocular diseases, *Doc. Ophthalmol. 43*(2): 217, 1977.
114. S. J. Crews, Chloroquine retinopathy with recovery in early stages, *Lancet 2*:436, 1964.
115. J. R. Brinkley, E. L. Dubois, and S. J. Ryan, Longterm course of chloroquine retinopathy after cessation of medication, *Amer. J. Ophthalmol. 88*:1, 1979.
116. R. P. Burns, Delayed onset of chloroquine retinopathy, *New Engl. J. Med. 275*:693, 1966.
117. A. Scherbel, M. G. Mackenzie, J. E. Nousek, and M. Atdjian, Ocular lesions in rheumatoid arthritis and related disorders with particular reference to retinopathy, *New Engl. J. Med. 273*:360, 1965.
118. H. E. Hobbs, S. P. Eadie, and F. Sommerville, Ocular lesions after treatment with chloroquine, *Brit. J. Ophthalmol. 45*:284, 1961.
119. E. Okun, P. Gouras, H. Bernstein, and L. Von. Sallman, Chloroquine retinopathy, *Arch. Ophthalmol. 69*:59, 1963.
120. L. M. Proenza, J. M. Enoch, and A. Jampolsky, *Clinical Application of Visual Psychophysics*, Cambridge University Press, Cambridge, 1981.
121. U. Nylander, Ocular damage in chloroquine therapy, *Acts Ophthalmol. Suppl. (Kbh) 92*:5, 1967.
122. H. Voipio, Incidence of chloroquine retinopathy, *Acta Ophthalmol. (Kbh) 44*:349, 1966.
123. U. Nylander, Ocular damage in chloroquine therapy, *Acta Ophthalmol. (Kbh) 44*:335, 1966.
124. O. Carlberg, Three cases of chloroquine retinopathy. A follow-up investigation, *Acta Ophthalmol. (Kbh) 44*:367, 1966.
125. A. I. Friedmann, The early detection of chloroquine retinopathy with the Friedmann visual field analyzer, *Ophthalmologica (Suppl.) 158*:583, 1969.
126. S. J. Crews, Some adverse effects of drugs on the retina, In *Evaluation of Drug Effects on the Eye*, P. V. Pigott, Ed., F. J. Parsons Ltd., London, 1968.
127. R. E. Carr, P. Gouras, and R. D. Gunkel, Chloroquine retinopathy. Early detection by retinal threshold test, *Arch. Ophthalmol. 75*:171, 1966.
128. S. J. Crews, The prevention of drug-induced retinopathies, *Trans. Ophthalmol. Soc. U.K. 86*:63, 1966.
129. S. P. B. Percival and I. Meanock, Chloroquine: Ophthalmological safety and clinical assessment in rheumatoid arthritis, *Brit. Med. J. 3*:579, 1968.
130. D. Adlakha, S. J. Crews, A. C. I. Shearer, and E. L. Tonks, Electrodiagnosis in drug-induced disorders of the eye, *Trans. Ophthalmol. Soc. U.K. 87*:267, 1968.
131. J. Babel and E. Meyer, Frequence et prevention des lesions retiniénnes dues à la chloroquine, *Schweiz Med. Wochenschr. 95*:1125, 1965.
132. P. Henkind, Screening test for chloroquine retinopathy, *Lancet 2*:40, 1964.
133. T. P. Kearns and R. W. Hollenhorst, Chloroquine retinopathy: Evaluation by fluorescein fundus angiography, *Trans. Amer. Ophthalmol. Soc. 64*:217, 1966.
134. J. A. Osterhuis, T. N. Boen-Tan, Fluorescein fundus angiography in chloroquine retinopathy, *Ophthalmologica Suppl. 158*:615, 1969.
135. S. Rampon, F. Rouher, R. Alfiere, P. Sole, J. L. Bussiere, and B. Sauvezie, Surveillance ophtalmologique des malades traites par antipaludéens de synthèse: Mise en evidence electroretinographique de l'impregnation retiénne toxique, *Rev. Rhum. Mal. Osteoartic. 37*:233, 1970.
136. P. Henkind, R. E. Carr, and I. M. Siegel, Early chloroquine retinopathy: Clinical and functional findings, *Arch. Ophthalmol. 71*:157, 1964.

137. H. Kolb, Electro-oculogram findings in patients treated with animalarial drugs, *Brit. J. Ophthalmol. 49*:573, 1965.
138. J. D. Reagan and J. A. Parrish, *The Science of Photomedicine*, Plenum Press, New York, 1982.
139. R. A. Weale, Senile changes in visual acuity, *Trans. Ophthalmol. Soc. U.K. 95*:36, 1975.
140. K. Arundale, An investigation into the variation of human contrast sensitivity with age and ocular pathology, *Brit. J. Ophthalmol. 62*:213, 1978.
141. D. B. Archer, Fluorescein angiography, In *Heredity Retinal and Choroidal Diseases*, A. E. Krill, Ed., Harper & Row, Hagerstown, MD, 1972.
142. G. E. Korte, R. W. Bellhorn, and M. S. Burns, Ultrastructure of blood–retinal barrier permeability in rat phototoxic retinopathy, *Invest. Ophthalmol. Vis. Sci. 24*(7):962, 1983.
143. J. S. Glaser, P. J. Savino, K. D. Sumers, S. A. McDonald, and R. W. Knighton, The photostress recovery test in the clinical assessment of visual function, *Amer. J. Ophthalmol. 83*(2):255, 1977.
144. R. E. Carr, P. Henkind, N. Rothfield, and I. M. Siegel, Ocular toxicity of antimalarial drugs: Long term follow-up, *Amer. J. Ophthalmol. 66*(4):738, 1968.
145. J. S. Dobson and P. A. Davison, A new rapid test of contrast sensitivity function utilizing spatial band with equalization, *Invest. Ophthalmol. Vis. Sci. 19*(2):213, 1980.
146. H. Zwick, R. B. Bedell, and K. Bloom, Spectral and visual deficits associated with laser irradiation, *Mod. Probl. Ophthalmol. 13*:299, 1974.
147. W. H. Merigan, T. Pasternak, and D. Zehl, Spatial and temporal vision of macaques after central retinal lesions, *Invest. Ophthalmol. Vis. Sci. 21*:17, 1981.
148. G. B. Arden, The importance of measuring contrast sensitivity in cases of visual disturbance, *Brit. J. Ophthalmol. 62*:198, 1978.
149. I. Bodis-Wollner and J. M. Camisa, Contrast sensitivity measurement in clinical diagnosis, In *Neuro-Ophthalmology*, S. Lessell and J. T. W. Van Dalen, Eds., Excerpta Medica, Amsterdam, 1980.
150. D. Regan, J. Raymond, A. P. Ginsburg, and T. J. Murray, Contrast sensitivity, visual acuity and the discrimination of Snellen letters in multiple sclerosis, *Brain 104*:333, 1981.
151. G. B. Arden and J. Jacobson, A simple grating test for contrast sensitivity: Preliminary results indicate value in screening for glaucoma, *Invest. Ophthalmol. Vis. Sci. 17*(1):23, 1978.
152. I. Bodis-Wollner, Vulnerability of spatial frequency channels in cerebral lesions, *Nature 261*:309, 1976.
153. R. L. DeValois, Spatial timing of LGN and cortical cells in monkey visual system, In *Spatial Contrast*, H. DeSpekreijse and L. H. van der Tweel, Eds., North Holland, Amsterdam, 1977.
155. I. Bodis-Wollner, Visual acuity and contrast sensitivity in patients with cerebral lesions, *Science 178*:769, 1972.
156. A. Fiorentini and L. Maffei, Spatial contrast sensitivity in myopic subjects, *Vis. Res. 16*:437, 1976.
157. A. Sjostrand and L. Frisen, Contrast sensitivity in macular disease, *Acta Ophthalmol. 55*:507, 1977.
158. J. M. Enoch, Perimetric standards and perimetric glossary of the International Council of Ophthalmology, Hague, The Netherlands, 1979.
159. W. M. Hart, R. M. Burde, G. P. Johnston, and R. C. Drews, Static perimetry in chloroquine retinopathy, periforeal patterns of visual field depression, *Arch. Ophthalmol. 102*:377, 1984.
160. H. M. Burian and M. C. Fletcher, Visual functions in patients with pigmentary degeneration following the use of NP-207, *Arch. Ophthalmol. 60*:612, 1958.
161. N. Accornero, A. Berardelli, G. Cruccu, and M. Manfredi, Computerized video screen perimetry, *Arch. Ophthalmol. 102*:40, 1984.
162. T. K. Kuyk, The effects of stimulus size and retinal location on the increment threshold spectral sensitivity of humans, *Invest. Ophthalmol. Vis. Sci.* (Arvo Suppl.) 19:136, 1980.
163. C. Noorlander, J. J. Koenderink, R. J. den Ouden, and B. W. Edens, Sensitivity to spatiotemporal colour contrast in peripheral visual field, *Vis. Res. 23*:1, 1983.
164. G. L. Alexander, Diagnostic value of colour fields in neurosurgery, *Trans. Ophthalmol. Soc. U.K. 76*:235, 1956.

165. S. K. Bhargava and C. I. Phillips, Foveal spectral thresholds in tobacco amblyopia, *Acta Ophthalmol. (Kbh) 52*:67, 1974.
166. W. M. Hart, R. K. Harz, R. W. Hagen, and K. W. Clark, Color contrast perimetry, *Invest. Ophthalmol. Vis. Sci. 25*(4):400, 1984.
167. H. Krastel, W. Jaeger, and S. Braun, The contribution of spectral increment thresholds to the interpretation of color perimetry, *Dev. Ophthalmol. 9*:171, 1984.
168. M. A. Johnson and R. W. Massof, The effect of stimulus size on chromatic thresholds in the peripheral retina, *Doc. Ophthalmol. Proc. Ser. 33*:15, 1982.
169. H. G. Sperling and R. S. Harwerth, Red-green cone interactions and the increment threshold spectral sensitivity of primates, *Science 172*:180, 1971.
170. A. E. Krill, Evaluation of night vision: Dark adaptation, In *Hereditary Retinal and Choroidal Diseases*, Vol. 1, Harper & Row, Hagerstown, MD, 1972.
171. K. J. W. Craik and M. D. Vernon, The nature of dark adaptation, *Brit. J. Psychol. 32*:62, 1941.
172. P. Dieterle and E. Gordon, Standard curve and physiological limits of dark adaptation by means of the Goldmann–Weekers adaptometer, *Brit. J. Ophthalmol. 40*:652, 1956.
173. B. Drum, Cone threshold vs retinal eccentricity: Changes with dark adaptation, *Invest. Ophthalmol. Vis. Sci. 19*(4):432, 1980.
174. A. C. J. Shearer and B. E. Cooper, The assessment of retinal function by electrophysiological techniques, In *Evaluation of Drug Effects of the Eye*, P. V. Pigott, Ed., F. J. Parsons, London, 1968.
175. A. E. Krill, A. M. Potts, and C. E. Johanson, Chloroquine retinopathy. Investigation of a discrepancy between dark adaptation and the electroretinographic findings in advanced stages, *Amer. J. Ophthalmol. 71*:530, 1971.
176. M. M. Connel, B. J. Poley, and J. R. McFarlane, Chorioretinopathy associated with thioridazine therapy, *Arch. Ophthalmol. 71*:816, 1964.
177. G. Verriest and D. Laplasse, New data concerning the influence of ethyl alcohol on human visual thresholds, *Exp. Eye Res. 4*:95, 1965.
178. H. L. Evans and R. H. Garman, Scotopic vision as an indicator of neurotoxicity, In *Neurotoxicity of the Visual System*, W. H. Merigan and B. W. Weiss, Eds., Raven Press, New York, 1980.
179. J. Pokorney, V. C. Smith, G. Verriest, and A. J. L. G. Pinckers, *Congenital and Acquired Color Vision Defects*, Grune & Stratton, New York, 1979.
180. G. Verriest, Further studies on acquired deficiency of color discrimination, *J. Opt. Soc. Amer. 53*:185, 1963.
181. A. J. Adams, R. Rodic, R. Husted, and R. Stamper, Spectral sensitivity and color discrimination changes in glaucoma and glaucoma-suspect patients, *Invest. Ophthalmol. Vis. Sci. 23*(4):516, 1982.
182. R. G. Alken and U. Vitt, Effect of repeated therapeutic doses of digoxin on relative red and green sensitivity detected by Nagel's anomaloscope in healthy trichromats, *Dev. Ophthalmol. 9*:219, 1984.
183. R. G. Alken, Drug-induced color vision deficiencies from side effects to clinical pharmacology, *Doc. Ophthalmol. Proc. Ser. 33*:467, 1982.
184. H. E. Henkes, Photic injury to the retina, *Doc. Ophthalmol. 44*(1): 113, 1977.
185. P. Grutzner, Acquired color vision defects secondary to retinal drug toxicity, *Ophthalmologica (Suppl.) 158*:592, 1969.
186. D. Farnsworth, The Farnsworth Munsell 100-hue and dichotomous tests for color vision, *J. Opt. Soc. Amer. 33*(10):568, 1943.
187. R. W. Pickford, A practical anomaloscope for testing colour vision and colour blindness, *Brit. J. Physiol. Optics 14*:2, 1957.
188. R. Laskowski, Calibration, validation and population norms for the Pickford-Nicolson anomaloscope, *Brit. J. Physiol. Optics 26*:166, 1971.
189. H. E. Kolder, Application of electrodiagnostic methods in Ophthalmology, *Dev. Ophthalmol. 9*:29, 1984.
190. K. T. Brown, The electroretinogram: Its components and their origin, *Vis. Res. 8*:633, 1968.

191. A. E. Krill, Electroretinogram, In *Hereditary Retinal and Choroidal Diseases*, Harper & Row, Hagerstown, MD, 1972.
192. G. Niemeyer, Electrophysiological testing of the function of the vertebrate retina, *Pharmacol. Ther. 5*:593, 1979.
193. W. Straub, ERG in acute and chronic intoxications, In *The Clinical Value of Electroretinography*, J. Francois, Ed., Karger, Basel, 1968.
194. R. D. Weekley, A. M. Potts, J. Reboton, and R. H. May, Pigmentary retinopathy in patients receiving high doses of a new phenothiazine, *Arch. Ophthalmol. 64*:65, 1960.
195. F. Liverani and U. Schaippi, Electroretinography as an indicator of toxic retinopathy in dogs, *Pharmacol. Ther. 5*:599, 1979.
196. K. A. Hellner and K. K. Gauri, Electroretinogram and ocular toxicity, *Dev. Ophthalmol. 9*:213, 1984.
197. K. Hommer and A. von Lutzow, Influences of some drugs on the ERG of the isolated rabbit retina, *Excerpta Medica Int. Congr. Ser. 160*:35, 1968.
198. W. R. Biersdorf and D. A. Diller, Local electroretinogram in macular degenerations, *Amer. J. Ophthalmol. 68*:296 1969.
199. J. L. K. Bankes, The foveal electroretinogram, *Trans. Ophthalmol. Soc. U.K. 87*:249, 1967.
200. S. Sokol and D. Nadler, Simultaneous electroretinograms and visually evoked potentials from adult amblyopes in response to a pattern stimulus, *Invest. Ophthalmol. 18*(8):848, 1979.
201. A. Fiorentini, L. Maffei, M. Pirchio, D. Spinelli, and V. Porciatti, The ERG response to alternating gratings in patients with diseases of the peripheral pathway, *Invest. Ophthalmol. Vis. Sci. 21*(3):490, 1981.
202. P. Bobak, I. Bodis-Wollner, C. Harnois, L. Maffei, L. Mylin, S. Podos, and J. Thornton, Pattern electroretinograms and visual-evoked potentials in glaucoma and multiple sclerosis, *Amer. J. Ophthalmol. 96*(1):72, 1983.
203. C. Teping and A. Groneberg, Physiological basis and clinical application of pattern electroretinogram, *Dev. Ophthalmol. 9*:74, 1984.
204. G. B. Arden, A. Barrada, and J. H. Kelsey, New clinical test of retinal function based upon the standing potential of the eye, *Brit. J. Ophthalmol. 46*:449, 1962.
205. A. E. Krill, Electro-oculogram, In *Hereditary Retinal and Choroidal Diseases*, Vol. 1, Harper & Row, Hagerstown, MD, 1972.
206. W. K. Noell, Experimentally induced toxic effects on structure and function of visual cells and pigment epithelium, *Amer. J. Ophthalmol. 36*:103, 1953.
207. P. Gouras and R. D. Gunkel, The EOG in chloroquine and other retinopathies, *Arch. Ophthalmol. 70*:629, 1963.
208. G. H. M. van Lith, G. T. M. Mak, and H. Wijnands, Clinical importance of the electro-oculogram with special reference to the chloroquine retinopthy, *Bibl. Ophthalmol. 85*:2, 1976.
209. G. H. M. van Lith, Electro-ophthalmology and side-effects of drugs, *Doc. Ophthalmol. 44*(1): 19, 1977.
210. H. E. Henkes, G. H. M. van Lith, and L. R. Canta, Indomethacin retinopathy, *Amer. J. Ophthalmol. 73*:846, 1972.
211. S. J. Crews, Some aspects of retinal drug toxicity, *Ophthalmologica 158*:232, 1969.
212. A. E. Krill, Visually evoked responses, In *Heredity Retinal and Choroidal Diseases*, Vol. 1, Harper & Row, Hagerstown, MD, 1972.
213. H. van der Marel, G. Dagnelie, and H. Spekreijse, Pattern evoked potentials in awake rhesus monkeys, *Invest. Ophthalmol. Vis. Sci. 21*(3):457, 1981.
214. G. P. Cooper, D. A. Fox, W. E. Howell, R. D. Laurie, W. Tsang, and J. P. Lewkouski, Visual evoked responses in rats exposed to heavy metals, In *Neurotoxicity of the Visual System*, W. H. Merigan and B. Weiss, Eds., Raven Press, New York, 1980.
215. A. Keeney, Ophthalmoscopy, In *Ocular Examination, Basis and Technique*, A. Keeney, Ed., C. V. Mosby, St. Louis, 1970.

216. A. R. Rosenthal, H. Kolb, D. Bergsma, D. Huxsoll, and J. L. Hopkins, Chloroquine retinopathy in the rhesus monkey, *Invest. Ophthalmol. Vis. Sci. 17*(12):1158, 1978.
217. E. S. Rosen, *Fluorescence Photography of the Eye*, Butterworths, London, 1969.
218. M. R. Stein and C. C. Parker, Drug reaction following intravenous fluorescein, ARVO Abstracts, 1971.
219. I. Ben-Sira, D. Weinberger, J. Bodenheimer, and Y. Yassur, Clinical method for measurement of light backscattering from the *in-vivo* human lens, *Invest. Ophthalmol. Vis. Sci. 19*(4):435, 1980.
220. J. A. van Best, J. L. van Delft, and J. A. Oosterhuis, Comparison between *in-vivo* and in-vitro fluorophotometry in the rabbit eye, *Ophthalmol. Res. 15*:318, 1983.
221. F. Kayazawa, Ocular fluorophotometry using high S-N ratio fluorophotometer, *Ann. Ophthalmol. 16*(5):472, 1984.
222. N. T. Yu, J. F. R. Kuck, and C. C. Askren, Red fluorescence in older and brunescent human tissue, *Invest. Ophthalmol. 18*(12):1278, 1979.
223. K. Sasaki, T. Shibata, M. Fukada, and O. Hockwin, Changes of lens transparency with aging: A clinical study with human volunteers using a Scheimpflug camera, In *Ageing of the Lens*, F. Regnault, O. D. Hockwin, and Y. Courtois, Eds., Elsevier, Amsterdam, 1980.
224. V. Dragomirescu, O. Hockwin, E. Weiglin, and H. R. Koch, The Scheimpflug photography of the anterior eye segment, its application in cataract evaluation, In *Ageing of the Lens*, F. Regnault, O. Hockwin, and Y. Courtois, Eds., Elsevier, Amsterdam, 1980.
225. S. Lerman and J. M. Megaw, Transmission characteristics of commercially available sunglasses, *J. Toxicol. Cut. Ocular Toxicol. 2*(1):47, 1983.
226. E. M. Farber (Ed.), Current status of oral PUVA therapy for psoriasis, eye protection revisions, *J. Amer. Acad. Derm. 6*:851, 1982.

INDEX